Carlson Wade's

AMINO ACIDS

Book

FIRST THINGS FIRST

Our bodies need the right balance of amino acids to make complete proteins, the basic material (*protein* literally means "the first thing") of living tissue. Amino acids also have amazing special properties, serving as natural drugless remedies for a wide range of conditions, including tension, depression, high blood pressure, herpes, alcoholism and insomnia.

Yet the food we eat often contains too few of some of these vital substances, robbing us of the complete protein we need for health and healing. This authoritative guide to the fabulous world of amino acids tells you:

- what amino acids are and why you need them;
- best food sources—with easy-to-prepare recipes for delicious, healthful dishes;
- how to combine foods for complete protein;
- power protein foods and recommended supplements;
- how to combine exercise and amino acid intake for total fitness;

—and everything else you need to know to use amino acids for the health you want for the rest of your life!

Other Keats Books by the Author

Vitamins, Minerals and Other Supplements
Arthritis, Nutrition and Natural Therapy
Bee Pollen and Your Health
Fats, Oils and Cholesterol
Hypertension and Your Diet
Water and Your Health (with Allen E. Banik)
What's in It for You?
Lecithin Book
Propolis: Nature's Energizer
PMS

Carlson Wade's

AMINO ACIDS Book

Keats Publishing, Inc. New Canaan, Connecticut

Carlson Wade's Amino Acids Book is not intended as medical advice. Its intention is solely informational and educational. Please consult a medical or health professional should the need for one be warranted. Neither the author nor the publisher has authorized the use of their names or the use of any of the material contained in this book in connection with the sale, promotion or advertising of any product or apparatus. Any such use is strictly unauthorized and in violation of the rights of Carlson Wade and Keats Publishing, Inc.

Carlson Wade's Amino Acids Book

Pivot Original Health Edition published 1985

Library of Congress Catalog Number: 84-80809

ISBN: 0-87983-372-6

Printed in the United States of America

PIVOT ORIGINAL HEALTH BOOKS are published by

Keats Publishing, Inc.
27 Pine Street (Box 876)
New Canaan, Connecticut 06840

CONTENTS

TABLES

Carlson Wade's
AMINO ACIDS
Book

INTRODUCTION

A proper balance of amino acids can enrich your bloodstream, rejuvenate your skin, build up your immune system, and regenerate your digestive system. You need these life givers and life extenders in adequate supply in your daily food program. They hold the key not just to survival, but also to healthy self-rejuvenation.

You need protein for life, and the amino acids are the building blocks of protein. To help you enjoy the best of life and health, you need to give your body a special balance of these amino acids. Any imbalance, however slight, can create problems. A missing amino acid is like a missing building block. The entire structure may threaten to collapse because of a single weakness.

For example, you may enjoy a corn-based diet, but corn is deficient in tryptophan, and this deficiency can cause emotional disorders and insomnia. If you add grains, seeds or nuts to the corn, you will provide the tryptophan necessary for brain nourishment.

Amino acids are found in all protein foods; but they have to be "complete," i.e., in complementary balance, and in sufficient quantities and potencies so that they work together with equal vigor in building and maintaining health. Complete amino acids are commonly found in animal products such as meat, fish, eggs, poultry and dairy products, but a daily diet of these fatty and high-cholesterol foods might be unwise. Your goal is therefore to eat a variety of foods from animal and non-animal sources for balanced nourishment. With the proper combinations, you can obtain amino acids in proper balance from meatless sources, if you wish. In so doing, you will protect yourself against fat and cholesterol overload but you will still be adequately nourished.

Amino acids build and maintain all body tissues; they constitute an important part of enzymes, hormones and body fluids. They are responsible for regulating body functions such as growth and digestion. They help form antibodies to fight infection.

In this book, you will discover that protein has the power to give you extended youth. But you will also learn that amino acids are the life-giving components of protein and essential in your quest for optimal health.

1 PROTEIN—Essential for Life

To Understand Better how amino acids can give you vibrant health, you need to become acquainted with protein. The following overview of the importance of protein is based on material by Ruth M. Leverton, nutrition scientist with the Agricultural Research Service of the U.S. Department of Agriculture in Washington, D.C.[1]

You are looking at a superb package of proteins when you see yourself in a mirror. All that shows—muscles, skin, hair, nails, eyes—is protein tissue. Teeth contain a little protein. Most of what you do not see is protein, too—blood and lymph, heart and lungs, tendons and ligaments, brain and nerves, and all the rest of you. Genes, those mysterious controllers of heredity, are a particular kind of protein. Hormones, the chemical regulators of body processes, and enzymes, the spark plugs of chemical reactions, also are proteins.

According to Leverton, the fact that protein is necessary for life was first recorded in 1838 by a Dutch physician-turned-chemist, Gerrit Jan Mulder. He announced his conclusion from many investigations that all living plants and animals contain a certain substance without which life is impossible. He did not know what was in it, but he was sure it was vital. He named it *protein*, from a Greek word, *protos*, meaning "first." Scientists since then have discovered that there are hundreds of different kinds of protein—not just the one substance Mulder observed. They also learned that proteins are unique in that they contain the element nitrogen. All our foodstuffs—fats, starches, sugars and proteins—contain the elements carbon, hydrogen and oxygen in varying proportions. Because proteins contain them and also nitrogen, they have a special importance and power.

HOW PROTEINS ARE MADE

Proteins have to be made by living cells. They do not exist in air, like nitrogen or oxygen, or come directly from the sun, like energy. Most plants make their own protein by combining the nitrogen from nitrogen-containing materials in the soil with carbon dioxide from the air and with water. The energy they need for the process comes from the sun. Examples: Legumes, which include beans, peas and peanuts, can use the nitrogen directly from the air for combining with the other substances to make protein.

Animals—including humans—cannot use such simple raw materials for building proteins. We must get our proteins from plants and other animals. Once eaten, these proteins are digested into smaller units and rearranged to form the many special and distinct proteins we need.

Is there any difference between plant and animal protein? Although we sometimes hear plant proteins referred to as "inferior" to animal proteins, plants are the basic factor of proteins. All proteins come directly or indirectly from plants. We depend heavily on farm animals to convert plant proteins into animal proteins for us, but most animals, too, must have some animal protein supplied to them. The ruminant animals—cattle, sheep, goats—are an exception, because they can use the simple nitrogen-containing substance in young pasture grasses; the microorganisms in their paunches can make microbial proteins, which the animal can then digest and use.

HOW PROTEIN/AMINO ACIDS BUILD LIFE

To understand why protein is the very source of life, consider the following:

Next to water, protein is the most plentiful substance in your body. If all the water were squeezed out of you, about half your dry weight would be protein. About a third of the protein is in the muscle, about a fifth in the bone and cartilage and about a tenth in the skin. The rest is in the other tissues and body fluids. Bile and urine are the only fluids that do not contain protein.

There are several dozen proteins in the blood alone. One of the busiest is hemoglobin, which constantly transports oxygen from the lungs to the tissues and brings carbon dioxide back from the tissues to the lungs. Ninety-five percent of the hemoglobin molecule is protein; the other 5 percent contains iron.

Other proteins in the blood are defenders, for they give us the means of developing resistance and sometimes immunity to disease. Gamma globulin can also form antibodies, substances that can neutralize bacteria, viruses and other microorganisms. Different antibodies are specific for different diseases.

Once we have had a disease, like measles, and the antibody for measles has been formed, it stays in the blood, and we are not likely to have measles again. A vaccination, such as for poliomyelitis, introduces a tiny amount of the inactive or dead virus into the body to stimulate the blood to form the specific antibody needed for neutralizing the virus that causes poliomyelitis. The presence of an antibody in the blood may give the person immunity to the disease. At least it gives him a head start in fighting the virus, and the disease will be less severe. Gamma globulin also helps the scavenger cells—phagocytes—engulf disease microbes.

PROTEINS BOOST NUTRIENT REACTIONS

Proteins help in the exchange of nutrients between cells and the intercellular fluids, and between tissues, blood and lymph. When one has too little protein, the fluid balance of the body is upset, so that the tissues hold abnormal amounts of liquid and become swollen.

The proteins in the body tissues are not there as fixed, unchanging substances deposited for a lifetime of use. They are in a constant state of exchange. Some molecules or parts of molecules always are breaking down and others are being built as replacements. This exchange is a basic characteristic of living things; in the body it is referred to as the dynamic state of body constituents—the opposite of a static or fixed state.

This constant turnover explains why our diet must supply adequate protein daily even when we no longer need it for growth. The turnover of protein is faster within the cells of a tissue (intracellular) than in the substance between the cells (intercellular).

PROTEIN SUPPLIES ENERGY

Proteins, like starches, sugars and fats, can supply energy. One gram of protein will yield about 4 calories when it is combined with oxygen in the body. One ounce will give 115 calories, about the same amount as starches and sugars.

The body puts its need for energy above every other need. It will ignore the special functions of protein if it needs energy and no other source is available. This applies to protein coming into the body in food and to protein being withdrawn from the tissues. Either kind gets whisked through the liver to rid it of its nitrogen and then is oxidized for energy without having a chance to do any of the jobs it is especially designed to do. This protein-sparing action of carbohydrates means that starches and sugars, by supplying energy, conserve protein for its special functions.

PROTEINS BECOME AMINO ACIDS

Amino acids are the chemical units of which proteins are made. The kinds and amounts of amino acids in a protein determine its nutritive or biological value.

The amino acid composition of animal muscle, milk and egg is similar, though not identical, to the amino acid composition of human tissues. Because these animal proteins can supply all of the amino acids in about the same proportions in which they are needed by the body, they are rated as having a high nutritive value.

The proteins from fruits, vegetables, grains and nuts supply important amounts of many amino acids, but they do not supply as good an assortment as animal proteins do. Their nutritive value is therefore lower. The proteins from some of the legumes—especially soybeans and chickpeas—are almost as good as those from animal sources.

To have the nutritive value of the mixture of proteins in our diets rate high requires only that a portion of the protein come from animal sources.

The body constantly uses materials for maintenance, regardless of the supply. It operates best when the supply of materials from food is generous and regular, but it does not stop functioning immediately when the food fails to supply what is needed. It mobilizes material from its tissues to meet these needs as long as that supply will last.

Suppose that the diet does not furnish enough protein for the body's daily operating and repair needs. The first thing the body does is to draw on some of its own tissue protein to supply this daily wear-and-tear quota. As a result, the operating and repair needs are met, and the normal kinds and amounts of end products of protein metabolism leave the body.

How much protein does the body need? The protein requirement depends on how fast the body is growing and how large it is. The faster the body is growing, the more protein it needs for building. The larger the mass of living tissue, the more protein it must have for maintenance and repair. A child grows faster during the first year than at any other time in his life. His second fastest growing period is during adolescence. His total need increases as he gets bigger, because there is more and more tissue to keep supplied and replenished with protein.

There are variations in the needs for men and women according to age and for certain specific needs, such as lactation. A rule of thumb would be one gram of protein for each kilogram (2.2 pounds) of weight. Therefore, if you are at normal weight, divide the figure by one-half and you have the approximate amount of protein you need daily.

RECOMMENDED DIETARY ALLOWANCE

Specific figures designed for the maintenance of good nutrition are set by the Food and Nutrition Board of the National Academy of Sciences-National Research Council.[2] Minimum amounts of protein needed on a daily basis are as follows:

Category	*Age*	*Weight*	*Daily Protein In Grams*
Children	1–3	29 lbs.	23
	4–6	44 lbs.	30
	7–10	62 lbs.	34
Males	11–14	99 lbs.	45
	15–18	145 lbs.	56
	19–22	154 lbs.	56
	23–50	154 lbs.	56
	51 +	154 lbs.	56
Females	11–14	101 lbs.	46
	15–18	120 lbs.	46
	19–22	120 lbs.	44
	23–50	120 lbs.	44
	51 +	120 lbs.	44
Lactating females			+30

Proteins come from two basic sources, animals and plants. Animal sources include lean meat, poultry, fish, seafoods, eggs, milk and milk products, yogurt and cheese. Plant sources include dried beans and peas, peanut butter and nuts, cereals, breads and pasta.

According to Malden C. Nesheim, Ph.D., professor of nutrition at Cornell University, dietary proteins vary in amino acid composition and in their ability to provide the essential amino acids needed by the body for protein synthesis.[3] These differences in protein quality affect the quantity of protein required.

Less is required of a protein with an amino acid composition similar to that of human body tissue (for example, meat) than of a protein containing some essential amino acids at levels that are very low relative to body protein (for example, wheat). Protein of animal origin (meat, milk, cheese, eggs) generally has an amino acid composition that more closely resembles human body tissue proteins than proteins found in plants.

Animal studies have shown that the protein in wheat is only about 30 percent as effective as the protein from milk. Note, though, that in a mixed diet proteins from plant sources are used quite effectively by the body because they are complementary.

A relative abundance of a certain essential amino acid present in some plant proteins makes up for the deficiency in others. For instance, milk protein with a cereal grain, or corn with dried beans will provide a more "complete" protein mixture.

Animal products are not the only way to consume protein. Protein-rich vegetables, such as legumes, will provide considerable amounts. One cup of kidney beans equals the protein of a two-ounce pork chop or serving of fish.

It is important to have a balanced daily intake of amino acids from protein. In health terms, a continuous, normal level of protein synthesis requires the continuous availability of the entire array of necessary amino acids. The removal of even one essential amino acid from the diet leads rather rapidly to a lower level of protein synthesis in the body, according to Dr. Nesheim.

FROM PROTEIN TO AMINO ACIDS

The protein you eat does not just remain as such in your digestive system. Instead, it is transformed. Each link in a long chain of protein molecules is an amino acid, which is the reason amino acids have been called the "building blocks" of protein. The chemical link from one amino acid to another is called a peptide bond. Food proteins are split during digestion into their constituent amino acids. These are then utilized in the synthesis of body protein for tissue growth and maintenance.

Your body can synthesize many of the amino acids from intermediates of carbohydrate metabolism and an appropriate source of nitrogen that can be transferred to the amino acid precursors by the process of transamination. However, your body cannot synthesize eight or nine of these amino acids which are therefore dietary essentials. These "essential" amino acids are lysine, valine, isoleucine, leucine, threonine, tryptophan, methionine, phenylalanine and histidine. Two other amino acids can only be derived from amino acids which are dietary essentials . . . tyrosine from phenylalanine and cystine from methionine.

2 GETTING ENOUGH AMINO ACIDS?

YOUR BODY does not immediately assimilate protein as such. Your digestive system breaks it down into simpler compounds—the amino acids—which are then utilized as "building blocks" for your entire body. Both digestive and intestinal enzymes break down protein into usable amino acids. Then hormones leap into action, combine with digestive enzymes and seize their needed amino acid. One will be used to feed your red blood cells, another will strengthen your fingernails, another will feed your skin and hair, and another is immediately seized by your bloodstream to be used for the repair of worn-out and damaged brain capillaries.

Edmund Sigurd Nasset, Ph.D., professor of physiology at the School of Medicine and Dentistry of the University of Rochester, author of *Food and You*, explains it this way:

"All proteins consist of very large molecules, which in turn are made up of various combinations of smaller building blocks called amino acids. More than twenty different amino acids are known and each one contains at least one atom of nitrogen that can be utilized in the body. Ten of the amino acids have certain peculiarities of structure which the body is unable to synthesize from simpler substances. These are referred to as the *essential* amino acids and must be supplied pre-formed in the food."[4]

Opinions vary as to whether the "essential" amino acids number eight or nine or ten, but as a general guideline, here is a listing of ten important amino acids that your body cannot manufacture but must obtain from food. Check your symptoms. See if your body is trying to tell you that you are deficient in one or more amino acids.

1. **VALINE.** Needed to spark mental vigor, muscular coordination, smooth functioning of the nervous system. *Deficiency symptoms:* Nervous reactions, fingernail biting, poor mental health, inability to sleep properly.
2. **LYSINE.** Most body growth factors are dependent upon lysine. This amino acid nourishes the blood and helps form antibodies to fight disease by building immunity. Newer studies in nutrition reveal that lysine can correct the imbalances leading to herpes, the virus that causes painful mouth sores and fever blisters. Some research suggests that lysine also plays a role in reducing dental caries or tooth decay. *Deficiency symptoms:* Visual disorders such as "red spider webs" in the eyes may be traced to poor lysine intake. Chronic tiredness and fatigue are other possible symptoms.
3. **TRYPTOPHAN.** For rich, red blood, youthful skin and healthy hair, you need this amino acid. It helps in utilization of the B-complex vitamin group and promotes better digestion. It has been demonstrated to be an "emotional stabilizer" to help combat depression. A natural source of niacin, it has been found to aid sleep. In a rather complex process,

tryptophan is converted into serotonin, a neurotransmitter that transfers nerve impulses from one cell to another. Serotonin is said to be responsible for normal sleep because it reduces the electrical activity of the brain. Tryptophan also is being studied for its effects on stress reduction, depression and alcoholism. *Deficiency symptoms:* Inability to sleep, poor skin coloring, brittle fingernails, prematurely aging skin, indigestion.

4. **METHIONINE.** The cells of the liver and kidneys need this amino acid for regeneration. It also helps relieve arthritic-rheumatic disorders; strengthens hair follicles; helps remove poisonous wastes from the liver and protects against necrosis or destruction of the delicate tissues and capillaries of this vital organ. *Deficiency symptoms:* Poor skin tone, hair loss, toxic waste buildup, malfunctioning of the liver.
5. **CYSTINE.** This amino acid supplies an appreciable amount of insulin needed by the pancreas for assimilation of sugars and starches. It is being studied for its role in building proteins such as hair, as well as its ability to help destroy damaging chemicals in the body such as acetaldehyde and other chemicals produced by alcohol and tobacco smoke. It has been used as a detoxicant agent against heavy metals such as mercury and cadmium and as a protection for heavy drinkers and smokers against acetaldehyde poisoning. *Deficiency symptoms:* Poor skin tone, hair loss, toxic waste buildup, malfunctioning of the liver.
6. **PHENYLALANINE.** Your thyroid gland demands this amino acid for stimulation so it can secrete the iodine-rich hormone thyroxine. This hormone sparks healthy nerve activity and helps provide mental

equilibrium. Current research indicates that it acts as a powerful analgesic or pain reliever for many chronic conditions. These include low back pain, headaches, arthritis, neuralgias and sports injuries. Other research suggests that it has an anti-depressive effect, which is not surprising since chronic pain situations provoke concurrent depression. It is believed to help suppress the appetite and improve mental alertness and energy levels. It is used by the brain to manufacture norepinephrine and dopamine, which act as neurotransmitters, dispatching signals between the nerve cells and the brain. These are classified as excitatory transmitters, which function to improve your learning and memory. *Deficiency symptoms:* Emotional upset; poor vascular health; some eye disorders; runaway appetite and weight gain.

7. **ARGININE.** Males especially need arginine, since seminal fluids contain as much as 80 percent of this substance. Arginine also helps detoxify poisonous wastes and filter out toxic substances. It is believed to stimulate the release of the human growth hormone (HGH) by the pituitary gland in the brain. This hormone helps the body burn up fat and build firm muscles. *Deficiency symptoms:* Infertility, particularly in the male; overweight; buildup of free radicals or wastes in the bloodstream; premature aging.

8. **GLUTAMIC ACID.** Brain health is influenced by this amino acid and it helps correct personality disorders. In some situations, the intelligence and personality of the individual, together with overall mental and physical equilibrium, respond remarkably to glutamic acid. Some of the most stubborn and obnoxious individuals have gone through a com-

plete change of personality (for the better, that is) when given supplementary glutamic acid. *Deficiency symptoms:* Oldsters and others who are grouchy and cantankerous may become more cheerful.

9. **HISTIDINE.** The auditory nerve needed for good hearing is stimulated and "fed" by this amino acid. *Deficiency symptoms:* Poor hearing or deafness. Injuries to the nerve cells of the hearing mechanism that make it difficult to distinguish words.
10. **THREONINE.** Needed to help the digestive and intestinal tracts work much more smoothly. Also helps improve the assimilation and absorption of nutrients for overall health. *Deficiency symptoms:* Indigestion, acid upset, stomach disorders, malabsorption, weak assimilation and general malnourishment.

By a strange quirk of nature, amino acids are not interchangeable. Each one has a specific purpose of its own. One cannot be substituted for another. You need ALL the amino acids for overall superior mental and physical health.

FOUR BASIC FUNCTIONS OF AMINO ACIDS

The amino acids have four functions:

1. They are building blocks used to make billions of body tissues and cells. These "blocks" are the raw materials required to synthesize growth and repair in *all* parts of your body.

2. Thousands of enzymes needed for digestion of foods are created by amino acids. These wonder-workers also help produce hormones and glandular secretions. Reproductive powers as well as virility and fertility depend upon a good amino acid supply.

3. Because of their structural properties, amino acids are vital for the functioning of the bloodstream.

4. Amino acids give you energy. As soon as the specific amine has been chopped off from the amino acid, the molecule that remains has a lot of carbon but little nitrogen. If this molecular amino acid is not needed immediately the body changes it into glucose and glycogen. In these forms, it is stored in the liver to be used in situations when more energy is required. You can take out "energy insurance" in the form of stored up amino acids.[5]

HIGH BIOLOGICAL VALUE

Nutritionists with the National Livestock and Meat Board are aware of more than just the value of protein; they feel that this nutrient should have something extra. "Proteins that contain all of the essential amino acids in proportions most useful to the body are described as having high biological value. The protein quality of a food is rated accordingly.

"In general, animal foods—meat, poultry, fish, eggs, milk and cheese—contain top quality protein. This means that these foods supply necessary amounts of all essential amino acids. Grain foods and many vegetables and fruits contain some of the important amino acids but not all in sufficient quantity to make them as beneficial in the diet. For the most efficient utilization of plant protein, an animal protein should be included in the same meal. For this reason, it is well to include a serving from the meat group or dairy group at every meal.

"All essential amino acids must be in the bloodstream at the same time for the body to take care of its job of tissue growth and repair."

The Board points out that in extreme cases of protein deficiency, there may be poor muscle tone and posture, lowered resistance to disease, premature aging, anemia, stunted growth in children, tissue degeneration, edema and slow recovery from illness or surgery."[6]

Many therapeutic diets include amino acids as a means of boosting recovery and healing. The Board lists these seven as examples of the healing power of protein-amino acids:

1. *Premature infants*. High quality protein and iron aid in the formation of hemoglobin, thus helping to improve an anemic condition found frequently in premature infants.

2. *Diabetes*. Some physicians put diabetic patients on a diet containing as much as 100 grams of protein daily. Its satisfying quality helps cut down on the danger of "going off" the prescribed diet. Also, there is considerably less glucose produced by protein (58 percent) than by carbohydrate (100 percent), an important consideration in the diet of the diabetic.

3. *Iron deficiency anemia*. Nutritional anemia, characterized by a deficiency in the quality or quantity of blood, occurs when there is an insufficient amount (or poor utilization) of iron or protein or both in the diet for a considerable period of time. A diet high in both these nutrients becomes important in the treatment of this disease. It has been estimated that one in every eight young women is anemic. Proper diet can help prevent, as well as cure, this disease.

4. *Preoperative and postoperative patients*. The preoperative patient who is in proper nutritional balance before surgery recuperates more rapidly than those who have not eaten proper levels of protein and other nutrients. Many body materials are lost following surgery. A high-protein diet (100 to 125 grams) is used with beneficial results to restore lost body protein. Because of its protective action on the liver, protein helps mitigate the toxic effect of anesthesia and minimizes the

chance of postoperative shock. Adequate protein also helps prevent infection and aids in wound healing. Where there is severe blood loss, extra amounts of iron, as well as protein, become doubly important. B-complex vitamins, potassium, zinc, copper, phosphorus and magnesium are also important in the diet of the postoperative patient.

5. *Peptic ulcer*. Most authorities recommend a higher-protein diet for ulcer patients than for those in good health. It is sometimes necessary to increase the amount of iron. In addition to providing needed nourishment, protein foods help neutralize the acid in the stomach.

6. *Disease of the liver*. Since one of the chief objectives in the treatment of diseases of the liver (jaundice, infectious hepatitis, cirrhosis) is to aid in the regeneration of liver tissue, a high-amino acid, high-carbohydrate and moderate-fat diet is usually prescribed. Vitamin supplementation is also useful.

7. *Kidney disease*. A high protein diet has proven to be effective in the treatment of chronic nephritis, where it is important to rebuild body tissues. Since building can occur only when all amino acids are present in the bloodstream, it is important that foods containing high-quality protein, such as meat, are included at every meal. When the condition is accompanied by edema, sodium is restricted in the diet.

ARE THEY ACID FOODS?

Amino acids are not "acids" in the usual sense of the word. Most amino acids are neutral, but they have been chemically classified as amino acids or *bases* as a means of helping scientists decipher the means by which they get into the body and across membranes to perform their function. Bases means alkaline, which is the opposite of acidic.

AMINO ACIDS ARE NEEDED DAILY

To enjoy better health and a longer and more youthful life, you should plan to nourish your body and mind with a daily intake of amino acids. This importance is underscored by John D. Palombo, M.L.S., M.S., and George L. Blackburn, M.D., of the nutrition metabolism laboratory of New England Deaconess Hospital/ Harvard Medical School in Boston. Here is their explanation: "Next to water, protein is frequently considered the quintessential nutrient, since it is the most essential constituent of cell metabolic components. The protein content of the adult human body is approximately 12 kilograms, half of which is contained in the body cell mass.

"The predominant protein mass of the body that requires dietary protein for replenishment of indispensable amino acids is the lean body mass (i.e., fat-free muscle, viscera and connective tissue).

"The protein content of the human body is constantly undergoing synthesis and breakdown. The rates and directions of these processes in the various tissues are dependent upon many factors, including the age and physical condition of the subject, the dietary intake of calories and protein, and physiological and pathological occurrences, i.e., starvation, infection, stress and disease."[7]

The liver and muscle proteins, which constitute a major part of the body cell mass, are synthesized from dietary amino acids and amino acids provided from tissue protein breakdown. Surplus amino acids are used to produce carbon skeleton molecules or are broken down into carbon dioxide, water, energy and nitrogen-based radicals for excretion or utilization in other compounds. In the liver, the amino group is converted to urea and subsequently excreted in the urine.

Protein nitrogen is lost continuously through the urine, feces and skin, as well as through the hair and nails. The daily protein supply for tissue protein synthesis must therefore be continuously replenished.

Protein is called "the most important ingredient in our diet" by Donald L. Donohugh, M.D., of the American College of Physicians and author of *The Middle Years*. "It occurs in both plant and animal tissues, and its molecules are made up of various combinations of the twenty amino acids that are its building blocks. Only eight amino acids cannot be manufactured by the adult human body, so these eight are termed essential amino acids—it is *essential* they be provided in our diet."[8]

Dr. Donohugh explains further, "Proteins can be categorized as complete, partially complete, or incomplete, depending upon whether they have enough of the essential amino acids in proper proportion, along with a good supply of the others, to maintain the body in a healthy state.

"The complete proteins are mostly of animal origin—eggs, milk, meat, poultry and fish. Wheat germ and dried yeast approach the animal sources in terms of quality. Partially complete proteins are those such as beans and peas that will allow life to continue, but not the body to grow, heal well, or respond to stress effectively. The incomplete proteins are those that will not even sustain life."

HOW PROTEINS BECOME AMINO ACIDS

This process is described by Dr. Donohugh in his highly acclaimed book. Digestion breaks down the proteins we consume into their constituent amino acids. These are then absorbed and carried to cells in the body, where they are made into other proteins or burned for energy. If protein is used as a source of energy, each gram produces 4 calories. Some special cells in the body make proteins that are required to sustain not just themselves, but the body as a whole.

When any proteins are constructed, all of the necessary amino acids must be present together at the same time. Therefore, a complete protein or a complementary combination of partially complete proteins should be eaten at least every day, if not at every meal.

Dr. Donohugh says, "It has always seemed to me that a simple numerical or color code on packages would help greatly in the selection and serving of complementary proteins, but this is yet to be." He feels that "if you have four servings per day of protein, half from an animal source and half from legumes such as beans or peas, you will have more than the recommended amount, and in proper balance."

Amino acids are therefore of prime importance in your quest for adequate nourishment and good health of body and mind. Had your amino acids today?

3 HOW AMINO ACIDS CAN REBUILD YOUR LIFE

JUST WHAT are amino acids? You know they are the building blocks that give you life and protection from, not to mention healing of, many illnesses. But in a chemist's laboratory, just what are these protein end products?

An overview is offered by agricultural food and nutrition scientist Ruth M. Leverton in *Food: 1959 Yearbook of Agriculture*.[9] In brief, most of the amino acids exist in two forms or patterns. A unique feature of the amino acids is that the arrangement of the carbon, hydrogen, oxygen and nitrogen atoms (and others if they are present) can exist in two patterns. One pattern is the mirror image of the other, like your left and your right hand. Nature makes only the left-hand pattern, called the L-form of amino acids, and this is the pattern found in all our foods. In general, your body can use only the L-form.

When chemists make amino acids synthetically in the laboratory, they come out with a mixture of equal parts of the left-hand and the right-hand (D-form) patterns. The body cannot use the right-hand pattern of an amino acid except as a source of carbon and nitrogen, which it may build into the L-form of certain amino acids.

The amino group makes it possible for an amino acid to act like a base (also called an alkali), while the acid group makes it possible for it to act like an acid. This dual action is one of the special characteristics of amino acids. Whether they can act as an acid or a base depends on which is needed at the moment to keep the acid-base balance of the body, especially of the blood, within normal limits. Proteins are often referred to as buffers because of this ability, through their amino acids, to protect the body against sudden or great changes in its acid-base relationships.

It is through their amino and their acid groups that amino acids are joined together to make proteins. The acid group of one molecule of amino acid reacts with the amino group of another just as any acid and alkali react together.

A molecule of water is formed and travels off, leaving the nitrogen of one amino acid joined to the carbon of the next amino acid. Such a joining is called a peptide linkage. A protein is a group of amino acids held together by peptide linkages. Specific enzymes in the gastrointestinal tract attract the peptide linkages when proteins are digested. First, the protein is separated into many clumps of amino acids. Then the clumps are separated further into single amino acids, which are absorbed from the intestine and carried by the blood to the liver.

But the amino acids do not stay single for long. As soon as they leave the liver and are carried by the blood to different tissues, they are reassembled into the special combinations that make the proteins to replace cell material that has worn out, to add a tissue which needs to grow or to make some enzyme or hormone or other active compound.

It is remarkable how the normal body has unerring

accuracy in assembling amino acids into the vital substances needed in every location. If any amino acids are left over, they cannot be stored in the body for use at a later time. They are returned instead to the liver and stripped of their amino groups in a process called deaminization. The nitrogen leaves the body chiefly as urea through the urine, but the carbon, hydrogen and oxygen fragments that are left can be used to provide energy. If the energy is not needed immediately, the fragments can be converted to fat and stored for use at a later time.

You need *all* of the amino acids if they are going to have an opportunity to rebuild your life and give you abundant health. The pioneering nutritionist Adelle Davis was ahead of her time when she pointed to the risks of a deficiency in just one amino acid. In her classic book *Let's Eat Right to Keep Fit*, she urged total amino acid intake for total health. She wrote:

"A tremendous amount of research is being done with both animals and humans to find the specific symptoms of ill health that occur when certain amino acids are lacking. For example, when the diet of animals or babies lacks tryptophan, methionine, or isoleucine, the liver cannot produce the blood proteins albumin and globulin (antibodies), and urine can no longer be collected normally; swelling, known as edema, and susceptibility to infections result.

"Methionine has been found to be particularly deficient in the diets of children with chronic rheumatic fever and of women suffering from the toxemia of pregnancy.

"In animals a lack of tryptophan or methionine causes the hair to fall out; a lack of histidine, phenylalanine, or any one of several other amino acids causes the eyes to become bloodshot and/or cataracts to form.

"An undersupply of arginine causes animals to become sterile and brings about a decrease in the formation and mobility of sperm in men, whereas too little tryptophan causes the animals' testicles to degenerate (atrophy) or females to lose their young.

"A deficiency of methionine allows fat to be retained in the liver of both animals and humans.

"Only future research can give an understanding of the role each amino acid plays in building and maintaining the body. It is known, however, that all the amino acids are used together and that taking one or two alone can never build health."[10]

AMINO ACIDS ARE WONDER WORKERS

Remember the bit of folk wisdom that drinking a glass of warm milk before you go to bed will help you sleep better? Like other folk remedies, this one has a scientific basis. Milk contains tryptophan, a natural sleep-inducing amino acid that helps you relax so that you will be able to drift off to sleep without the use of chemicals.

"In Japan, arginine water is sold as a health drink," notes Yoon Sang Cho-Chung, M.D., Ph.D., chief of the cellular biochemistry section of the laboratory of pathophysiology at the National Cancer Institute in Bethesda, Maryland. And arginine is an important amino acid, which Dr. Cho-Chung has found "helps straighten out" breast cancer cells. With the use of arginine, "the breast cancer cells become flatter and enlarged—more like normal cells."

Dr. Cho-Chung believes that arginine is involved in the body's production of cyclic AMP (adenosine monophosphate, a compound occurring in cells and important in processes requiring the transfer of energy).

She feels that cancerous cells contain less cyclic AMP than healthy ones and found that when she treated breast tumor cells with cyclic AMP, she was able to halt their growth. She believes that just treating the cells with arginine will stop the cancer growth but she has found best results when using a combination of cyclic AMP and arginine. "The tumor-inhibiting effect is enhanced, even if we use an amount of cyclic AMP that is ineffective when used alone. We have found that we can lower the dose of cyclic AMP if we use arginine."[11]

In another report, Dr. Cho-Chung told of being able to stop the growth of cancer cells in test rats after just two days of treatment. The cancer remained stable for as long as the treatment continued. She concluded that "cyclic AMP and arginine may have therapeutic potential for breast cancer in humans."[12]

Another amino acid, phenylalanine, has been used to treat Parkinson's disease. In this condition, there are symptoms of tremor, rigidity and spontaneous movement. It is most pronounced in resting limbs, interfering with such actions as holding a cup. The patient has an expressionless face, an unmodulated voice, and an increasing tendency to stoop. A shuffling run is needed to maintain balance.

In a study conducted at the National University of Buenos Aires, phenylalanine showed remarkable abilities to alleviate symptoms. All but two out of fifteen patients with the disability and depression of Parkinsonism could "walk away" from their symptoms after four weeks of phenylalanine supplementation. This amino acid reportedly reduced rigidity in all fifteen patients; it improved speech in eleven out of twelve. But doctors report that the most "spectacular" improvements were noted in the emotional outlook of the patients. In fourteen cases, depression was lifted.[13]

Researchers at Queen Charlotte's Maternity Hospital in London found that some patients with depressive illness were unable to produce enough of three substances called trace amines. Since the body makes those amines from phenylalanine, they theorized that supplements of this amino acid would help ease general depression.[14] In another study, a team of Boston and Cambridge doctors treated a thirty-year-old woman with drugs for relief of severe depression. Her condition worsened. When they gave her supplements of tyrosine, her condition "improved markedly" after only two weeks. "She said she felt better than she had in years and showed striking improvement in mood, self-esteem, sleep, energy level, anxiety and somatic [physical] complaints."

To determine whether the amino acid really did the trick, a psychiatrist tried a placebo [dummy pill] instead of tyrosine. This was done without the knowledge of the woman or the doctor who treated her. Results? The researchers report that "within one week of placebo substitution, her depressive symptoms began to return." After eighteen days, she was more depressed than before she took the amino acid. She was given tyrosine again and her depression was again completely alleviated.[15]

Another case study described two patients who suffered such a severe degree of depression that they did not respond to conventional chemotherapy, according to the New York Psychopharmacologic Institute. To be free of symptoms, the two depressed patients had to take powerful amphetamines or "speed." They were given a special form of tyrosine before breakfast; after two weeks, the first patient was so improved that amphetamines were dropped completely, and the second was able to reduce his dosage by two-thirds.

Epilepsy is any one of a group of disorders of brain function characterized by recurrent attacks that have a sudden onset. The patient is in a state of clouded awareness and afterward may have no recollection of the event. The problem of epilepsy appears to be partially solved with the use of taurine, believed to be one of the most abundant amino acids in the body, particularly in the brain and muscles.

Ryan Huxtable, Ph.D., pharmacology professor at the Tucson-based University of Arizona Health Sciences Center, found that in animal studies, the injection of only a minute amount of taurine into the brain gives up to eighteen days' protection against epileptic seizures. In people, the use of taurine as an anticonvulsant agent has met with varying success. "However, there are two direct applications of taurine for people. First, I believe taurine should be added to infants' formula. Human mother's milk is rich in taurine but formula is not. It's possible that the lack of taurine in formula may be causing abnormalities in some infants that we just haven't yet connected with a taurine deficiency.

"Second, I believe taurine should be given as a supplement to people who have been stressed by surgery or by myocardial infarction. The amount of taurine in the affected tissues plummets after surgery or infarction. And we know, for example, that taurine is important in helping the heart to beat properly."

Dr. Huxtable adds that there are Japanese researchers searching for the relationship of taurine to blood pressure. It reportedly has been able to reduce pressure in experimental animals that are genetically predisposed to elevated readings.[17]

A virus-caused infection, herpes Type One is characterized by painful cold sores, usually on the lips. Inflammation of the skin and collections of small blisters are typical symptoms. It is believed that seven out of ten adults in the world will have this affliction at some time. How can an amino acid counterattack herpes? The answer comes from Richard S. Griffith, M.D., and colleagues at the Wishard Memorial Hospital, Indiana University School of Medicine in Indianapolis, who devoted three years to a study of this problem. Dr. Griffith treated forty-five herpes patients whose ages ranged from four to sixty by giving them lysine supplements. Dosages were from 312 to 1000 milligrams daily. According to the researchers, "The pain disappeared abruptly, overnight, in virtually every instance." Furthermore, while the patients took this amino acid, no new sores appeared, and existing sores healed more speedily than in the past.

When should you take lysine supplements? Around the time you feel a herpes outbreak approaching. Dr. Griffith explains that some patients are able to predict such occurrences and should be prepared. Herpes may crop up during menstruation, after exposure to sunlight, or after eating a particular food such as chocolate or nuts. Since everybody is different, the outbreaks can be traced to different causes. Know what you are sensitive to, and consider lysine as a protective and healing amino acid.[18]

WHEN DO YOU NEED MORE AMINO ACIDS?

Nutrition scientist Dr. W. R. Beisel, in an article in the *American Journal of Clinical Nutrition*, suggests that the need for protein may be greatly increased during episodes of stress, infection and injury. Alterations in protein metabolism are generally aimed at mobilizing skeletal muscle proteins to ensure an adequate supply of material for energy production and protein biosynthesis. This process is triggered by hormonal changes that result in an overall negative nitrogen balance.[19]

The degree of nitrogen loss depends upon the severity of infection or injury, notes nutritional scientist N. S. Scrimshaw, in the *American Journal of Clinical Nutrition*. Obligatory losses may approach the equivalent of 0.9 grams a day. Therefore, prolonged infection and injury without adequate nutritional support can cause serious protein or amino acid malnutrition. This would suggest boosting protein intake during a recovery period at the rate of about 1 gram a day or more, depending upon individual need, as recommended by the attending physician.[20]

AMINO ACIDS SOOTHE STRESS

How can we ease or reduce the mental burden that is responsible for stress? One answer appears to be brewer's yeast, a good source of mind-soothing B-complex vitamins but also amino acids, according to Ronald J. Amen, Ph.D., a nutritional physiologist in Anaheim, California. What's the connection?

"It has to do with the kinds of amino acids that make up the protein found in yeast," explains Dr. Amen. "You see, brewer's yeast protein is high in the so-called branched-chain amino acids known as leucine, isoleucine and valine. These amino acids are important combatants of the physiology of stress.

"Whether it's emotional or physical stress you are under, the effects are the same," points out Dr. Amen. "Either way, during times of stress your body requires an excessive amount of energy. To get enough, it breaks down proteins and their component amino acids and uses them for that energy.

"But when branched-chain amino acids are in your system you are better able to combat that breakdown. That's because some of those branched-chain amino acids are used as an energy source, saving the other amino acids to do the job they were originally intended for—the building of new proteins. These special amino acids actually act as regulators of the body's protein-building machinery and are called into action especially during times of stress.

"The absence of the branched-chain amino acids in times of stress may result in a patient's becoming so nutritionally compromised that it is more difficult to fight off the illness. On the other hand, patients who receive high levels of branched-chain amino acids appear to heal more quickly. Also, it is thought that those three amino acids are required in a specific ratio to obtain maximum results. Brewer's yeast meets those requirements perfectly."[21]

Brewer's yeast is an excellent food supplement with the important amino acids for overall health improvement. You need to take just one or two tablespoonsful each day. Blenderize it in a fruit or vegetable juice. Or just sprinkle on wholegrain cereal, hot or cold. Add it to baked goods of almost any kind. (It is also available in tablet form.) It is the great way to boost your amino acid intake.

AMINO ACIDS FOR THE DIETER

When you cut down on caloric intake, your body must use its internal reserves of fat and protein primarily to meet energy requirements. On a short-term diet program, your body adapts to the caloric deprivation by using muscle protein and adipocyte (cellular) fat as its primary source of calories. But if you diet longer than two weeks, your body undergoes extensive protein losses.

According to Drs. J. P. Flatt and George L. Blackburn, in the *American Journal of Clinical Nutrition*, protein replenishment is vital to ensure survival. If you are on a reducing program that advocates long-term unsupplemented fasting, you may jeopardize your health through excessive loss of functional tissue amino acids, vitamins and minerals. Dietary protein as a supplement will help attenuate nitrogen loss and achieve nitrogen equilibrium.[22]

It is important to obtain adequate amino acids on any type of reducing diet, since in recent years there have been some sixty deaths related to the use of very low-calorie, low-protein diets. Of these, sixteen were due to the development of cardiac arrhythmias and arrest in obese but otherwise healthy women who were reducing their body weight by strict adherence to a fasting regimen supplemented daily with small (3 to 5 ounces) quantities of predigested liquid protein products.

The cause of these cardiac arrhythmias could not be determined by investigators from the Food and Drug Administration, Center for Disease Control or an *ad hoc* committee of scientific experts convened by the FDA. While there is support for the hypothesis that these deaths were due to severe caloric restriction and starvation, the most likely explanation is that deficiencies of certain essential minerals (e.g., potassium, magnesium and phosphate) developed, along with serious amino acid deprivation. Several of these sixteen deaths occurred during carbohydrate refeeding, a period in which significant metabolic changes from the starvation state occur (e.g., increased sodium and water retention, increased insulin production, and altered intracellular enzyme activity and breakdown rates of muscle and adipose tissue).[23]

When following any reducing diet under medical

supervision, recognize that your body is undergoing a certain amount of stress, and this calls for more protein. With a balance of amino acids to meet the needs of bodily changes during reducing, you should be able to enjoy good health as the pounds come off.

4 ESSENTIAL VS. NONESSENTIAL AMINO ACIDS

WHAT ARE the basic differences between essential and nonessential amino acids?

Agricultural scientist Ruth M. Leverton, in *Food: 1959 Yearbook of Agriculture*, offers this basic interpretation: "Amino acids form the alphabet of the proteins. They have the same relation to proteins that letters have to words. At least twenty-two different letters make up the amino acid alphabet, and combinations of the same or different amino acids make a great variety of proteins. Not all of the amino acids are present in each protein, but there are many, many more amino acids in a protein than there are letters in any word."[24]

Leverton points out that eighteen different amino acids commonly occur in our food supply. Some are more important to us than others. The body can manufacture many of them from the materials supplied by protein and other substances in our food.

There are eight amino acids that the body must have but that it cannot make from any other substances. Our food must supply them completely formed and ready for use. They are valine, lysine, threonine, leucine, isoleucine, tryptophan, phenylalanine and methionine. They are called the essential or indispensable amino acids because it is essential to have them supplied ready-made.

Other amino acids are essential to life and health, too, but if our food does not provide any or enough of them, the body can make them from the raw materials supplied by the food. They are called nonessential or dispensable amino acids in reference to the fact that it is not essential for food to furnish them ready-made. They are glycine, tyrosine, cystine, cysteine, alanine, serine, glutamic acid, aspartic acid, arginine, histidine, proline and thyroxine.

The presence in a protein of all the essential amino acids in significant amounts and in proportions fairly similar to those found in body proteins classifies it as a complete protein—meaning that it could supply completely the needs of the body for these amino acids.

The proportions in which the essential amino acids are required are as important as the amounts. Apparently the body wants these amino acids to be available from food in about the same proportions each time for use in maintenance, repair and even growth. Meat, fish, poultry, eggs, milk, cheese and a few special legumes contain complete proteins. (Gelatin is the only food from an animal source that does not meet the specifications. It contains almost no tryptophan and has very small amounts of threonine, methionine and isoleucine.)

Often the proteins in grains, nuts, fruit and vegetables are classed as partially complete or incomplete because the proportionate amount of one or more of the essential amino acids is low, or because the concentration of all of the amino acids is too low to be helpful in meeting the body's needs. Grains, nuts and legumes are more concentrated sources of amino acids than fruit and vegetables.

Ruth Leverton carefully emphasizes that "timing, or the distribution of the amino acids among our daily meals, is a factor to consider in meeting our protein needs. In building body proteins, whether for growth or replacement, the body uses all of the essential amino acids plus the nonessential ones."

Suppose you take in an inadequate balance. When incomplete proteins are metabolized, they supply the body with enough of some of the essential amino acids but inadequate amounts of others. The amino acids that are supplied, however, are not used unless the other essential amino acids are present from other food sources. Instead, they are oxidized, and the nitrogen portion is excreted. The amino acids cannot be stored for use at a later time when other acids become available.

If the intake of any one of the essential amino acids is too small to meet the body's need, none of the other essential acids being fed can be used for growth or maintenance of tissue. They will be deaminized, and the nitrogen will be excreted. Also, the body will be in negative nitrogen balance because it has to use some of its own tissue protein as a source of the needed amino acids.

HOW MANY AMINO ACIDS DO YOU NEED?

The daily requirements of amino acids for nitrogen equilibrium range from about 0.25 gram of tryptophan to as much as 1 gram of leucine, methionine and phenylalanine. Intermediate amounts of the others are also needed. These amounts actually are very small compared with the amounts we eat in our protein foods every day.

The amino acids phenylalanine and methionine each have a special helper in a related but nonessential amino acid. Tyrosine can help phenylalanine so well that about three-fourths of the phenylalanine requirement can be met by tyrosine. There are still some functions that only phenylalanine can fulfill, but if there is only a limited supply of phenylalanine, it can be saved for those special functions if there is plenty of tyrosine.

Cystine is the helper for methionine and is the only other common and plentiful amino acid that contains sulfur. At least three-fourths of the methionine requirement can be met by cystine. Because of this high degree of interchangeability, we are likely to refer to the requirement for total sulfur-containing amino acids, realizing, of course, that at least a small amount of methionine must be present.

Almost any diet that includes a variety of everyday foods supplies generous amounts of all the essential and non-essential amino acids. When foods with proteins of high nutritive value (which means they have a high content of essential amino acids in good proportions) are included in the diet regularly, a person need not be concerned about the adequacy of his amino acids intake. It is possible, however, that an oversupply of one amino acid may reduce the utilization of another amino acid so that a deficiency will occur. Also, an excess of one amino acid may increase the requirement of another acid. The high leucine content of corn, for example, may increase the requirement for isoleucine.

How can you be sure you are receiving a balance of these amino acids in amounts that are helpful? Sometimes this can be done by careful combinations of foods. One way is to use a relatively small amount of a protein from an animal source to improve the quality of a protein from a plant source. If animal proteins are not used, different plant proteins must be combined so as to supplement each other. In some instances the main cereal grain can be supplemented with another grain—such as supplementing corn with wheat to improve the amount and proportion of tryptophan. Legumes, too, may be used as supplements. To achieve the goal of a good protein supply for everyone, adjustments may have to be made in some countries in the kinds of crops grown, in the methods of food processing and preparation, and in the food habits of the people.

HOW TO GET TOP-NOTCH BIOLOGICAL VALUE FROM YOUR AMINO ACIDS

In their book *Nutrition For Today*, Drs. Roslyn Alfin-Slater and Lilla Aftergood of the University of California make it clear that "all of the amino acids in the proper amounts are required for protein synthesis. Tissue proteins differ in their amino acid composition and in the amino acid ratios. If a particular dietary protein supplies all of the amino acids needed for the formation of tissue protein, it is considered to be of high biological value and is known as a complete protein. If any amino acids are missing from a protein food (especially if these are essential amino acids) this 'incomplete' protein is assigned a lower biological value.

"A diet selected from foods in the four food groups contributes mixtures of proteins of varying biological value; some are complete and others incomplete with regard to their amino acid composition. However, two incomplete proteins lacking different amino acids may provide a complete protein mixture when both are ingested simultaneously."[25]

For example, the proteins contained in tortillas and in beans are incomplete. However, the protein of tortillas lacks amino acids that are present in beans and vice versa; together, these two foods provide a complete amino acid mixture. This is known as mutual supplementation.

Another method of raising an incomplete protein with a low biological value to a higher rating is to supplement it with the amino acid(s) it lacks. The addition of the amino acid lysine, and in some cases methionine and/or tryptophan, to bread and cereal proteins (which are low in these amino acids) is an example of this type of fortification. The proteins of some species of nuts are often deficient in lysine and methionine and nuts are not considered to contain high-quality proteins.

In general, the proteins of animal products tend to be more complete and of higher biological value than those of vegetable products, which are incomplete. Agronomists and food technologists have been experimenting with improving the amino acid composition of vegetable proteins and, by a process of selection, high-lysine corn is now available.

Drs. Alfin-Slater and Aftergood note that "it has been established that for protein synthesis to proceed efficiently it is important to have all the amino acids present within a relatively short time (two hours). Therefore, an incomplete protein eaten at breakfast cannot be supplemented by another incomplete protein eaten at lunch."

If an essential amino acid is missing from the diet, protein synthesis is impaired and the amino acids are not used for tissue protein synthesis. As a result, there is an increased excretion of nitrogen and a negative nitrogen balance results. Continued ingestion of an inadequate protein results in serious decreases in tissue proteins and in plasma proteins as well.

Plasma proteins are used to transport many nutrients to various tissues of the body. If these are in short supply, tissues become depleted of their nutrient reserves. Also, since enzymes are composed of protein, a decrease in enzyme formation and activity may result that would affect the metabolism of the organism as a whole.

Drs. Alfin-Slater and Aftergood explain that "cooking protein foods not only makes them more palatable but also improves their digestion and absorption. In many cases, the biological value of proteins may be enhanced, since heating may increase the availability of some amino acids that originally may have been inaccessible to digestive enzymes. However, excess heat may make some of the amino acids unavailable by forming linkages that digestive enzymes cannot split. This is what may occur as a result of overheating during processing."

They also point out that other benefits accrue from the use of these amino acids. In addition to building new tissues during the growth process or after a wasting illness, proteins are required for the maintenance of existing tissues. A dynamic equilibrium exists between the proteins in the cells and the amino acids resulting from digestion of protein in foods.

They sum up by saying that proteins have important regulatory functions in the body. When the protein level in blood is low, extra water is drawn into the tissues and edema (excess water in tissues) results; this condition can be reversed by the ingestion of high-protein foods. In addition, proteins enter into systems that regulate the acid-base balance in the body and help keep the acidity of body fluids at the required level. The acidity of the body must be kept within narrow ranges to support life. Proteins are also constituents of important hormones and enzymes that regulate all types of metabolic processes in the body.

To reap the benefits of amino acids, they all should

be available in proper balance at the same time, whether essential or nonessential. This provides greater biological value, according to Dr. Edmund Sigurd Nasset, author of *Food and You*, who writes:

"During the normal digestion of a meal, amino acids are absorbed nearly as fast as they can be liberated from the proteins by the digestive enzymes. They are also absorbed at approximately equal rates, so that the blood, as long as digestion continues, is provided with a steady stream of amino acids representative of the constant mixture found in the gut. This is important because all the essential amino acids must be present in the blood at the same time if the body is to use them to the greatest advantage. That requirement is met by the cannibalistic device of mixing body proteins with food proteins in the alimentary tract before they are digested and absorbed. This does not mean that the body can continue indefinitely to thrive on poor-quality proteins. Eventually the body becomes so depleted of its own protein reserves that it is unable to furnish a satisfactory amino acid mixture for absorption—a fact that underscores again the nutritional importance of consuming a varied diet. When several proteins are mixed together in a meal, the body will not have any difficulty in obtaining the amino acid mixture that is optimal for health."[26]

Adelle Davis in *Let's Eat Right to Keep Fit* has a good suggestion for ensuring adequate protein intake: "Of all proteins available, the most concentrated and least expensive are brewer's yeast, powdered skim milk, wheat germ, soy flour and cottonseed flour. The use of these foods makes it possible to obtain protein on an extremely limited budget and can change a diet low in protein to one high in protein with little thought or effort. To obtain too little protein is a mark of carelessness or ignorance; to obtain too much is foolish and expensive; to obtain an adequate amount is to stay young for your years!"[27]

SPECIFIC FUNCTIONS OF AMINO ACIDS

In their valuable *Foods and Nutrition Encyclopedia*, Ensminger, Ensminger, Konlande and Robson offer this definition: "An amino acid is nonessential if its carbon skeleton can be formed in the body, and if an amino group can be transferred to it from some compound available, a process called transamination." They add that "aside from protein formation, some of the amino acids participate in other specific metabolic reactions and demonstrate some interesting characteristics." They cite:

Arginine. It participates in the formation of the final metabolic product of nitrogen metabolism—urea—in the liver. This process is known as the urea cycle.

Cysteine and methionine. These are the principal sources of sulfur in the diet. Sulfur is necessary for the formation of coenzyme A and taurine in the body. Methionine can convert to cysteine, but not vice versa. Methionine is also needed to form such compounds as epinephrine, acetylcholine and creatine.

Glutamic acid. Involved in creating gamma-aminobutyric acid, a substance needed for the nervous system.

Glycine. The simplest of the amino acids, it conjugates with a variety of substances, thus allowing their excretion in the bile or urine.

Histidine. A powerful blood vessel dilator; it is involved in allergic reactions and inflammation. Histidine also stimulates the secretion of both pepsin and acid by the stomach.

Lysine. This amino acid provides structural components for the synthesis of carnitine, which stimulates fatty acid synthesis within the cell. Lysine is apt to be deficient in vegetarian diets.

Phenylalanine and tyrosine. The body can convert phenylalanine to tyrosine, but the reverse reaction does not occur. In normal individuals almost all of the phenylalanine not used in protein synthesis is converted to tyrosine. Tyrosine becomes the parent compound for the manufacture of the hormones norepinephrine and epinephrine by the adrenal medulla (the inner part of the adrenal glands, the small triangular glands lying in front and on top of each of the two kidneys), and of the hormones thyroxine and triiodothyronine by the thyroid gland (which lies in the base of the neck).

Tryptophan. Used to manufacture a substance called serotonin, which helps the body relax and enjoy a better sleep cycle.[28]

Perhaps it is a misnomer to say that amino acids are either essential or nonessential. The fact is they are ALL needed in proper balance or in doctor-recommended specific supplements to help give you better health and a longer life.

5 HOW TO STAY YOUNG WITH AMINO ACIDS

WANT TO protect yourself against the restricting illnesses of so-called old age? You need to fortify your body with reserves of amino acids to help build resistance to the corrosive effects of day-to-day living in our polluted, overcrowded and stress-filled environment.

Why do you age? According to one scientist, Dr. Leo Szilard, random errors short of actual chromosome changes or mutations occur in certain processes within the cells with the passage of time and cause aging. These are dubbed "aging hits" by this scientist.[29] He further explains that aging is caused by errors in the system of protein synthesis and amino acid availability.[30]

This theory is upheld by another scientist, Dr. Leslie Orgel, who writes in the *Proceedings of the National Academy of Sciences* that cellular deterioration most likely caused by imbalance or deficiencies of specific amino acids is a prime reason for aging. A malfunctioning of protein in its transformation to amino acids can be a basic cause of the aging process.[31]

Donald L. Donohugh, M.D. of the American College of Physicians, author of *The Middle Years*, advances another interesting theory on the cause of aging as related to cellular deterioration.

"Throughout our lives, many cells in the body divide in the same way; some do continually, such as the cells lining the gastrointestinal tract and the blood cells. Others do so intermittently in order to repair damage—in the liver, kidneys and bones. Others, nerve and muscle cells, for instance, never divide. Whenever such division takes place, there is an increased chance of an error, or mutation.

"Mutations in body cells are called somatic to distinguish them from germ cell mutations. The rate of somatic mutations was first observed to increase with radiation, then found to occur at a regular rate over time as well. This latter observation resulted in the somatic mutation theory of aging, which holds that those cells that are so altered will no longer function properly, that in fact they will harm the host or produce substances that do and, by accumulation, cause aging."[32]

This brings us back to the "aging hits" theory, which imples a constant rate of errors occurring over a period of time. Like drops of water wearing away a stone, these "aging hits" can slowly drain away the very essence of life.

How to use amino acids as a means of protecting against cellular deterioration, the forerunner of aging?

THE YOUTH-PROLONGING PROPERTIES OF PROTEIN

Hans J. Kugler, Ph.D., author of *Slowing Down the Aging Process*, looks to protein/amino acids as a means of helping to forestall old age. He lists protein as one of the essential anti-aging nutrients (along with carbohydrates, fats, vitamins, enzymes and minerals). "They all play an important role in cell metabolism and the aging process."[33]

Kugler then explains the mechanism of the amino acids, which are of course derived from protein, and we can see how these powerful nutrients can build resistance to aging and help you become immune to many common and uncommon disorders:

A protein contains the elements carbon, hydrogen, oxygen, nitrogen and sometimes sulfur or other elements. When these elements combine in certain different ratios, they form the basic building blocks of proteins, the amino acids. Our body can synthesize some of these amino acids; those it cannot synthesize in sufficient quantities are called "essential" amino acids and must be supplied.

When different amino acids are combined in varying percentages and frequencies, the body produces certain proteins. That is why it is important that if you take protein preparations, you make sure that these preparations contain the essential amino acids, and in the right percentages.

"When you take in protein, it is first hydrolized (broken down) into its amino acids. The amino acids are then transported through the bloodstream to your cells, where they can be used to make the proteins that characterize the cell's molecular structure." (Significantly, amino acids containing sulfur play an important role in slowing down the aging process.)

"Proteins, then, are fundamental to every living material; they provide every cell in the body with its basic composition. If your body lacks proteins, it will take apart the least important body proteins in order to get the amino acids required for more important processes. But if we are talking about keeping our system in top shape, is there any protein that can be called less important?"

In effect, then, amino acids come to the rescue of

deteriorating cells; they nourish and strengthen them to help forestall or slow down the aging process.

"There are thousands of different proteins in our body," Kugler explains, "and each one requires the right amounts of essential amino acids. A decrease in the efficiency of your organs in old age is directly related to decreased and faulty protein synthesis in your body. It is therefore of the utmost importance to supply your body at all times with good proteins."

If you do not nourish yourself with ALL needed amino acids, you run the risk of these five aging-related reactions:

1. The maintenance of all organs and body processes will be hampered.
2. Thinking and memory processes will weaken.
3. Several necessary hormones will not be formed, since they depend upon proteins for satisfactory synthesis.
4. The correct enzymes cannot be formed because enzymes are basically proteins.
5. The formation of antibodies and other defenses that protect us from bacteria and other toxic materials will be seriously hampered because they are made of proteins.

Hans Kugler sums up as follows: "Making a protein is just impossible for your body if the necessary amino acids are not available. It's like asking a manufacturer to make machine parts out of iron only; if the man doesn't have any iron, he just can't do it no matter what you promise him or how hard you push him."

Amino acids are essential contributors to the continuous cell-feeding process, which is described by noted nutrition authority Adelle Davis in her classic *Let's Eat Right to Keep Fit:* "As your cells use amino acids, the supply is replenished from the breakdown of stored protein.

"As long as your diet is adequate, the amount of amino acids in your blood is kept relatively constant. If you ignore your health to the extent of eating insufficient protein, the stored protein is quickly exhausted. From that time on, the less important body tissues are destroyed to free amino acids needed to rebuild more vital structures. Such a process may go on month after month or even year after year. Your body continues to function after a fashion. Unseen abnormalities set in because blood proteins, hormones, enzymes and antibodies can no longer be formed in amounts needed. Muscles lose tone; wrinkles appear; aging creeps on; and you, my dear, are going to pot."[34]

In her highly acclaimed work *Let's Get Well,* Adelle Davis offers the following amino acid program, based on the protein content of foods:

"Foods supplying complete proteins of the highest biological value are, in descending order: eggs; milk and milk products; liver and other glandular meats; muscle meats, fish, and fowl; yeast, wheat germ, soy flour and a few nuts.

"Because the proteins in cereal grains, legumes and most nuts lack some of the essential amino acids, they are of value as proteins only when eaten with eggs, milk or meat or with each other so that all essential acids are furnished. The most rapid recovery from illness occurs if protein is supplied largely by eggs, milk, cheese and glandular meats."[35]

In your quest for an antidote to aging with the use of amino acids, you need to remember the importance of balance. Adelle Davis cautions that if an essential amino acid is lacking or a non-essential one is supplied in excess, a dangerous imbalance may occur, causing large amounts of the essential amino acids to be lost in the urine. She emphasizes that it is almost impossible to meet the protein needs of an ill individual unless a reasonably palatable drink is taken prepared with fresh milk fortified with foods supplying all the amino acids. Because yogurt and acidophilus milk are largely digested during the culturing process, they are particularly well tolerated by ill and elderly persons. When prepared at home and fortified with non-instant powdered milk, they can contain twice the amount of protein, calcium and vitamin B2 at a fraction of the cost of the commercial products; hence an electric yogurt maker—not of glass—is a good investment.

Cell-rejuvenating amino acids may be increased in the daily diet by fortifying such foods as cereals, hotcakes, waffles, muffins and cookies with wheat germ, soy flour and non-instant powdered milk. According to Adelle Davis, "instant powdered milk is excellent to add to fresh milk for drinking, but it makes foods gummy when used in cooking. Instantized foods, like puffed wheat compared with the natural product, are bulky. One-half cup of non-instant powdered milk, available at health food stores, equals a quart of fresh milk, compared with 1⅓ cups of the instant variety. Any form of powdered milk, however, has been somewhat harmed by heat and should not be substituted for fresh milk."

To get your supply of protein/amino acids, a simple plan appears in *ConsumerViews:*[36]

"A small serving of lean meat, poultry or fish (about 3 oz.) averages 20 grams of protein. Milk has 9 grams a cup; a 1 oz. slice of cheddar cheese, about 8; an egg, 6; bread, about 1.9 a slice.

"Food experts note that there is incomplete protein in grain foods and in some vegetables that is better utilized by the body when eaten together with a complete protein. So the generations of international cooks who combine small bits of meats, poultry, fish, cheese or eggs with beans, pastas and rice are right in terms of nutrition as well as flavor. Sandwiches, pastas with cheese or meat sauces, and cereal with milk are also efficient, nutritionally and budgetwise. (But be careful of fat-traps in rich sauces.)"

FOUR AMINO ACIDS THAT HELP CONTROL AGING

While your body will best respond to an anti-aging program that utilizes a balance of *all* amino acids, you may be weak or deficient in one or more of these potent factors and in need of a booster intake. In particular, scientists have found that four amino acids are of specific importance in the cell-rejuvenating process. You may want to fortify your body with either one or all of these amino acids; many health stores either have them available as supplements or can arrange to prepare them as per your order. Also check with your pharmacy.

These four amino acids contain sulfur; this is a potent substance that is able to protect you against radiation and pollution, the twin evils in our environment. These sulfur-containing amino acids help defuse free radicals, those cellular wastes and misshapen fragments that may be part of the underlying cause of aging. This amino acid group will neutralize the corrosive effects of poisonous wastes, increase the metabolism of protein and help metabolize vitamin C to build and rebuild your billions of body cells. The four amino acids are:

Methionine

This amino acid belongs to the lipotropic group (including inositol, betaine and choline) and functions to control the overload of fat in your liver. Methionine uses its sulfur content to support your liver's manufacture of lecithin, a valuable fat-fighting substance. An important key in rejuvenation may well be this protection against excess fat in your system.

Methionine acts as an antioxidant or deactivator of free radicals, which helps to slow down the aging process. It also has a unique ability to "chelate" or grab hold of metallic substances and heavy metals that might otherwise cause toxicity; these metals include dangerous lead, cadmium, and mercury, to name a few.

Taurine

Taurine is especially valuable in nourishing the brain cells; it works with choline to help maintain the neurotransmitters that promote thinking ability.

In some situations, brain malfunctions such as epilepsy can respond to taurine, offering a possible alternative to phenobarbital and other chemotherapeutic drugs. Some have dubbed taurine the "brain amino acid" and it does make sense to include it in your age-fighting program.

Cysteine and Cystine

These two are grouped together for a good biological reason. Cysteine is an unstable amino acid and is easily transformed into usable cystine. In most situations, it is referred to as cystine, the end product of the original cysteine.

Cystine uses its sulfur power in helping to mitigate the destructive effects of free radicals; it helps strengthen your powers of immunity. It especially works to repair the DNA-RNA components of your cells; these molecules need repair and regeneration to forestall and resist aging and extend your lifespan.

According to a report released by Dr. Eustace Barton-

Wright of London's Greenwich and District Hospitals, a combination of 100 milligrams of pantothenic acid and varying amounts of cystine has been used to help correct osteoarthritis and rheumatoid arthritis. When the cystine was removed, the results were negligible.[37]

A team of physicians at the Philpott Medical Center, now in St. Petersburg, Florida, headed by William Philpott, M.D., found that a combination of vitamin B6 and cystine could help initiate the rejuvenation reaction. "A majority of patients with chronic degenerative illnesses, whether physical or mental, have a vitamin B6 utilization disorder. The culprit in this vitamin B6 utilization problem seems to be an L-cystine deficiency." Because of this problem, body processes slow down and aging sets in. Dr. Philpott suggests that those who have a problem with the assimilation of vitamin B6 should consider taking 1.5 grams of L-cystine three times a day for a month, then twice a day. A doctor's supervision is strongly recommended.[38]

(*Note:* As explained earlier, amino acids exist in two forms, one being the mirror image of the other. The form used to make protein has been designated by scientists as the L-series. This explains the label of L-cystine [or any other amino acid]. Many distributors of amino acid products will make the label simpler by just dropping the L and using the name of the nutrient by itself.)

It is important to recognize that amino acids initiate responses and that they do this in a group and not singly. Except in situations when you may be biologically deficient in one or several amino acids, your body will respond better with the entire set, whether from foods or supplements.

At the same time, amino acids do not always act without other help, even as a group. They depend upon cofactors such as vitamins and minerals in order to become properly metabolized and able to function as precursors. Again, it is important to aim for a balance in your diet so that the metabolic process works in harmony.

6 HOW AMINO ACIDS BUILD IMMUNITY TO ILLNESS

AMINO ACIDS have a unique ability to neutralize and eliminate potentially destructive free radicals in your system, and thereby help build immunity to many illnesses.

The smallest killers in your body are not necessarily viruses, although these are dangerous enough. Rather, free radicals are the primary threat to your health. These are molecular fragments created during certain normal functions including the breakdown of fat and hydrogen peroxide (which white blood cells use to combat bacteria), and ordinary metabolism. Your body has a certain number of enzymes and amino acids that join together to neutralize and eliminate free radicals, but many still remain in your body, destroying cells, tissues and organs. This is the onset of the aging process. If free radicals are allowed to multiply, uncontrolled injury can extend to deterioration of the RNA-DNA components, and degenerative diseases will erupt. Free radicals can break down your immune system and cause you to become vulnerable to endless ailments.

Throughout your life, many body cells undergo division. Those cells lining the gastrointestinal tract, in the bloodstream, liver, kidneys, bones and everywhere else, will split off in order to repair injury. Then the cells have to join with others to reform and reshape themselves. This may give rise to the condition known as cross-linkage—namely, a bond between large amino acid molecules that normally are separate from each other. Fragments of these molecules break off and start to float in your system. These wastes from the cross-linkage condition are known as free radicals.

There is also a risk to RNA-DNA components. These are nucleic acids, compounds involved in amino acid synthesis. They are found in the nucleus of body cells. A problem here is that free radicals or wastes can cause injury to these compounds and thereby open the way to illness. The RNA-DNA compounds are the molecular basis of heredity; that is, they determine whether you will be well or ill, the length of your life and your level of immunity. The free radicals can so disturb these vital compounds that they can cause them to mutate, and this means a change in the number of chromosomes and genes that could spell breakdown in overall health.

Three causes of cellular deterioration are outlined by Hans Kugler in his *Slowing Down the Aging Process:* "The major factors that cause the destruction of cells are (1) the damage of the DNA, the central director of all cell functions; (2) an increase in large molecules due to cross-linking of smaller molecules; and (3) a constant decrease in the effectiveness of all cell functions due to 'stress.' Factors that wear out your cells and drive them to the point where they divide too early are faulty nutrition, smoking, lack of exercise, air pollution and emotional stress in general."[39]

This can trigger off the aging process, which Hans Kugler describes as a "constant loss of cells," and also the loss of natural immunity to many of life's debilitating illnesses. He believes that cell loss is due to five basic causes:

1. The efficiency of the cell is decreased because of toxic chemicals that interfere with metabolic processes; these include air pollution, smoking, alcohol, certain food preservatives and drugs.

2. Cross-linkage forms large molecules that clog up the entire system; such offenders include free radicals, oxidation products and metal ions.

3. The DNA in the nucleus is damaged by cross-linking and by free radicals.

4. A deficiency of the proper nutrients, such as amino acids, hampers the metabolic processes.

5. The cell wall and other cell parts are constantly damaged by free radicals; the latter are formed because of the effects of radiation, air pollution and faulty nutrition.

Kugler feels that these factors cause a breakdown in the immune system that leads to aging and even death.

Cellular destruction as a cause of premature aging and chromosome damage as a factor in lowered immunity are also problems recognized by Dr. Howard J. Curtis of the Brookhaven (N.Y.) National Laboratory. He has written numerous papers on the problem of chromosome aberrations that can open the way to illness and shorten the lifespan. He hopes that ways can be found to repair the damage and help improve the quality of life.[40]

AMINO ACIDS AS ANTIOXIDANTS

To understand how amino acids protect your body from the oxidation process that can bring about a breakdown in health, it is helpful to understand the corrosive dangers of these waste products.

Here is an explanation from *Health News & Review:* "Your cells' molecules are constantly undergoing biological or chemical reactions. During this process, free radicals are formed. These are highly volatile broken-off pieces of molecules that latch onto other molecules. In so doing, they damage amino acid molecules, enzymes, DNA or RNA (the material that determines the genetic code of each of your trillions of cells and is responsible for immunity to disease and protection against aging).

"Free radicals overcome the electrons of other molecules and claim them for their own use. The danger is that these free radicals can combine with fatty acids to form peroxides, which 'rust' the membranes of cells. A chain reaction is initiated and more free radicals are created. It has been found in laboratory tests that such a chain reaction may be the reason for failures of the body's immune system; infection-fighting lymphocytes actually attack the body's own tissues, and illness and aging soon follow.

"Shortly after announcement of the free radical theory of the cause of aging, gerontologist Dr. Johan Bjorksten of the Bjorksten Research Foundation in Madison, Wisconsin, found that a major probable cause of cellular disintegration was cross-linkage. He explained that free radicals could cause cross-linkage, the binding together of large molecules that should not be so connected. Dr. Bjorksten added that cross-linking two molecules is similar to handcuffing together two workers on an assembly line. As a result of this bondage, they would be handicapped or completely incapacitated. In biological terms, this reaction could cause systems and functions in your body to slow down, be rerouted or halt."[41]

How can antioxidants help combat free radicals? Oxidants are those wastes which are cell-destructive, including the free radicals. An antioxidant is a nutrient which prevents the formation of free radicals and oxidant wastes. It guards against rancidity in the cells, and helps detour the efforts of oxidants to destroy the cell membranes. It is also helpful in scrubbing away clinging wastes, and actually helping to wash your cells clean so that you become healthy and vigorous.

Amino acids head the list of antioxidants.

They act as oxygen carriers and work with vitamins to absorb and wash out the free radicals. In particular, the amino acids join with vitamin C to form two compounds inside the cells, 2,3-diketogulonic acid and dehydroascorbic acid, that are believed to have cleansing and protective reactions.

The sulfur-containing amino acids are singled out by Hans J. Kugler as having specific properties that help protect fats from oxidating and guard your cells against free radical invasion.[42] These are methionine, taurine, cysteine and cystine, mentioned earlier. They create corrective biochemical changes that help guard against oxidation and the subsequent penalty of premature aging. You may consider these sulfur-containing amino acids as free radical deactivators because they neutralize and wash out potentially damaging toxins.

AMINO ACIDS THAT BUILD YOUR IMMUNE SYSTEM

Nip the oxidative processes and cellular destruction in the bud with a variety of helpful amino acids. These include:

Cysteine. Able to detoxify your body, promote improved healing and boost resistance to disease.

Glutamine. An important amino acid, it cleanses nerve pathways so you can think better. May be considered an energizer for your brain.

Glutamic acid. Cleanses debris and washes your central nervous system so you are calmer.

Lysine. Helps build resistance to bacterial invasion, promotes a feeling of wellbeing; involved in the body's growth and immunosuppressive factors.

Methionine. Said to help cleanse the liver and kidneys, control cholesterol and wash out toxic wastes.

Phenylalanine. Gives important energy, wipes away gloom, helps soothe pain. Creates inner strength so that you have more resistance to common and uncommon disorders.

Taurine. Important for strengthening your brain wave patterns; cleanses away free radical wastes that might otherwise cause congestion that erupts as hypertension or stroke.

Tryptophan. Helps cleanse debris from your bloodstream; cleanses your skin and promotes healthier hair. Helpful in washing away tension and stress.

Tyrosine. Helps restore mental alertness, a better memory and more cheerful disposition.

Arginine and ornithine. They appear to work together in building your immune system. This pair helps control the aging factor and promotes better healing of wounds. In particular, they cause the release of a hormone from your pituitary gland which initiates the cell-washing process. They stimulate muscle growth and the burning of excess fat and wastes.

Glutathione. This is a naturally occuring triamino acid compound of cysteine, glutamic acid and glycine. In this form, it is believed to help neutralize and eliminate destructive free radicals.

Glycine. Necessary for the immune system, for balanced growth of white blood cells and for health of the thymus gland, spleen and bone marrow. This amino acid is a prime initiator of antibody action that helps fight off infectious bacteria. It is also involved in the manufacture of specific growth hormones that are needed to build strong immunity and anti-aging factors. Glycine is involved in muscle cell metabolism; that is, it is naturally converted into creatine, which acts as a store of high energy phosphate in muscle and serves to maintain adequate amounts of ATP (adenosine triphosphate), the source of energy for muscle contraction. This protects you against muscular fatigue.

Stress, infection and heat can cause increased nitrogen loss and a deficiency in amino acids, notes Dr. E. Consolazio in the *American Journal of Clinical Nutrition*. This calls for amino acid increase. Furthermore, this nutrition scientist notes that heavy work or intensive physical training exercises cause an increase in muscle mass; this condition also calls for more amino acids in order to meet these challenges.[43]

Dr. Denham Harman has reported in several medical journals that by using protective antioxidants in the diet of test animals, he could increase the lifespan of some (not all) by as much as 30 percent. Along with other scientists, he noted that somatic mutation—spontaneous errors—caused by division of cells, could also be a predisposing factor to aging. The use of antioxidants may be helpful in neutralizing free radicals so as to block or nullify cellular damage. The use of substances such as vitamin E and amino acids has been seen to be most helpful in building natural immunity.[44]

7 HEAL YOURSELF WITH AMINO ACIDS

SPECIFIC AMINO ACIDS are often used therapeutically in a variety of conditions. You may ask why you need supplements if you are already eating a nutritious diet. The answer is that amino acids are not stored in your body; you need a supply every single day. And as you have already read, your body cannot manufacture all of them, and must obtain them from food. The key here is to have a balanced diet of wholesome, natural foods that provide adequate amino acids, among other nutrients. But your body may have particular needs for specific amino acids that cannot be supplied by average foods. There may also be physiological conditions that inhibit formation of amino acids to the degree that you require for your particular condition. Supplements may be necessary to make up for any such deficiency.

Here is a rundown of the therapeutic uses of amino acids, either singly or in groups.

HEAL MENTAL DEPRESSION

A Canadian family was diagnosed as having a disease that brought on sudden severe mental depression along with related nervous symptoms. As reported in *Medical World News*, the patients' blood and brains showed very low levels of taurine because of an inherited intestinal disorder that blocked absorption of taurine. With the use of taurine supplements, there was hope for healing this condition.

INDUCE NATURAL SLEEP, PREVENT PAIN

For nearly twenty years, researchers at Tufts University (near Boston) have used tryptophan to help cure insomnia. They report amazing success in easing insomnia and depression with the use of this single amino acid.

Alice Kuhn Schwartz, Ph.D., co-author of *Somniquest*, herself a former insomniac, notes that you can overcome *initardia* (difficulty in falling asleep) by eating the proper foods in the right combination. What are these?

"First, foods that are high in the essential amino acid tryptophan. Eggs, almost any meat, certain fish such as salmon or bluefish, and dairy products, especially cottage cheese—all of those protein-rich foods have a hefty amount of tryptophan and this amino acid has been shown in the lab to increase drowsiness and help bring on sleep.

"But there is a second type of food you should eat as well. You see, in addition to tryptophan, there are many other amino acids. Once these different amino acids get into the bloodstream, they compete with each other for entry to the brain. For someone who wants to get to sleep, the trick is to give tryptophan a competitive edge."[45]

This maybe done by eating a carbohydrate food with the amino acid food. It "liberates tryptophan and gives it greater access in the brain." Alice Schwartz explains that tryptophan is hardly used by your brain without the availability of a carbohydrate at the same time. A carbohydrate-fat combination seems to work best.

"If you've been eating high-protein, high-tryptophan foods during the day, and you want to fall asleep at night, then it may help to eat some bread, have a banana, drink some grape or apple juice, have some figs or dates—all of those high-carbohydrate foods and many others will help activate tryptophan." The combination should be eaten within two to four hours before bedtime. This enables the food to reach its peak effect when you are ready to retire.

Suppose you wake up frequently throughout the night? This happens to many people over forty. What to do? Eat your carbohydrate immediately before turning off the lights. This helps your carbohydrates work with tryptophan after you have been asleep a few hours. What if you awaken in the middle of the night and cannot get back to sleep? "If you are still awake for a half hour, I recommend getting out of bed and performing some boring, routine task until you feel sleepy."

Tryptophan is sleep-inducing because when the tryptophan level goes up in the brain, so does that of serotonin, a neurotransmitter that tends to bring on a feeling of drowsiness and a desire for sleep.

Dr. Samuel Seltzer, a dentist at Temple University in Philadelphia, reports that he uses tryptophan to prevent chronic pain in dental patients. He gave this amino acid to thirty patients and found that there was a drop in the "pain rating" as well as an increased tolerance to pain.

Another essential amino acid, phenylalanine, has also been found to be an effective and all-natural pain reliever for migraine, whiplash, arthritis, sprains and low back pain.

Dr. Seymour Ehrenpreis, a pharmacologist at the University of Chicago, reported in *Pain Research and Therapy* that he used phenylalanine in clinical treatment of pain and found it was most useful. It appears to activate brain substances called endorphins, brain narcotics that occur naturally and relieve pain. Reportedly, this amino acid controls pain for long amounts of time so it does not need to be taken on a daily basis. It is not effective for transient pain or a sudden pain; but it is helpful for chronic pain.

HELP SOOTHE DEPRESSION

Tyrosine, another amino acid, is also much in the news as a means of treating depression. Drs. Richard Wurtman and Alan J. Gelenberg reported in the *Journal of Psychiatric Research* that they gave tyrosine to eleven depressed patients; of these, seven showed improvement. In five patients who were given a dummy or placebo pill, there was no improvement. But when these same five were later given tyrosine, three did improve.

In commenting on the depression-relieving effect of this amino acid, *Medical World News* adds: "Pending possible FDA action that could enable physicians to prescribe brain amine precursors (like tyrosine) for depression as they prescribe drugs, some doctors are dispensing tyrosine and tryptophan under Investigational New Drug permits, and others are sending patients to health food stores for the amino acids."

Drs. Wurtman and Gelenberg of Harvard Medical School are using funds from the National Institutes of Health to conduct a double-blind trial comparing the effects of an antidepressant drug, a placebo and tyrosine. The two researchers believe that certain kinds of depression could respond to a combination of tyrosine and tryptophan. They found that high-carbohydrate foods taken with these two amino acids helps facilitate their access to the brain.

KICK THE COCAINE HABIT

A combination of an antidepressant drug plus tyrosine and tryptophan helps cocaine addicts kick the habit, according to Dr. Jeffrey S. Rosecan of New York City. He reported to a World Congress of Psychiatry meeting in 1983 that twenty patients were given the drug plus one to two grams each of tyrosine in the morning and tryptophan before going to bed. Fourteen of the twenty stopped the use of cocaine entirely, while six decreased its use. Dr. Rosecan's objective is to show that the two amino acids, used in this way, help normalize the patients' "cocaine-deranged neurochemistry."

PREVENT HERPES SIMPLEX COLD SORES

Two researchers at Indiana University used lysine to cure cold sores, prescribing it in dosages ranging from 312 to 1200 milligrams. Nearly all of the 250 patients in the study responded; only two showed no improvement.

The researchers noted that lysine was able to accelerate recovery from a herpes attack. Pain vanished overnight, lesions stopped spreading and healing was rapid. To prevent recurrences, maintenance doses were taken, and worked for almost all the patients. When lysine therapy was halted, there was an immediate return of cold sores. The researchers found that if lysine therapy is begun when stinging pain signals a recurrence, the attack can be averted.

The preceding reports, appearing in *Health Foods Retailing*, offer hope for healing with one or more natural amino acids.[46]

HELP CORRECT FATIGUE

Your blood sugar governs your level of energy. If you boost blood sugar levels, you boost energy and correct fatigue. And this can be done with amino acids through the use of protein. According to Emrika Padus, author of *Woman's Encyclopedia of Health and Natural Living*, "If your energy seems especially pitiful in midafternoon, it could be your morning breakfast habit that's to blame. Think protein. And by that, we don't mean to advocate the traditional bacon-and-eggs breakfast that has become the mainstay of the American morning fare. Bacon is not only full of fat, it tends to be treated with cancer-causing nitrates, and eggs are often fried up with lots of grease. Consider more healthful protein options—items you can even eat on the run. How about a hard-boiled egg? Or a piece of cheese, or a cup of yogurt with fruit?"

Author Padus explains that the "coffee-and-a-donut routine doesn't solve the breakfast problem any better than no breakfast at all. Sugar in the morning sends your blood sugar soaring—but only temporarily. When it comes back down, it does so with a bang, dragging your energy down to the pits. In fact, one woman told us that she managed to beat the three o'clock slump simply by omitting sugar from her morning coffee."[47]

Basically, you need to keep your blood sugar level moderately high but *consistent* throughout the day. How? Say "no" to high-sugar and refined foods. Say "yes" to high protein/amino acid foods such as cottage cheese, lean meat and grain-legume combinations, and complex carbohydrates such as fresh fruit, vegetables and whole grains. Another tip, according to Emrica Padus, is the following: "Instead of the usual three gluttonous meals a day that tend to leave you stuffed and then famished later on, try to eat smaller but more frequent meals throughout the day."

With high amino acid foods and smaller meals, you should be able to have a stabilized blood sugar level that gives you plenty of energy throughout the day.

PROTEIN + ZINC = BETTER BONE HEALTH

Trauma is a generalized term to describe a variety of injuries of body and mind. If you are a victim of trauma, a combination of protein and zinc can help create faster healing, especially in bone disorders.

This phenomenon was noted in a research study by Augusta Askari, Ph.D., assistant professor in the department of surgery at the Medical College of Ohio, in Toledo. Together with her colleagues, she studied the zinc and nitrogen metabolism factor in injured rats. It was noted that trauma speeds up losses of both of these substances, at a time when they are especially needed. And remember, nitrogen is an end product of amino acids. In tests, it was found that rats with broken hind legs lost more zinc and nitrogen than those not injured. Askari comments: "Zinc is an essential trace element vital to life. It's part of RNA and DNA—it's in every single cell in the body. In trauma, it appears to be particularly important because of its role in wound healing and the formation of new protein."

Basically, trauma deprives the body of these substances just when they are needed to form new proteins, heal wounds and speed up recovery. With the use of zinc and amino acids that form nitrogen, there appears to be a better chance for prompt healing of bone disorders.[48]

BOOST SEXUAL DESIRES

Sexual vigor depends to a large degree on healthy glands, and glands require amino acids for proper functioning. In commenting on this biological fact, George Schwartz, M.D., associate professor of community, family and emergency medicine at the University of New Mexico Medical Center, author of *Food Power*, notes: "It is logical that our sex drive would go down when not enough protein or protein of the wrong amino acid type is eaten. [This refers to meatless protein or vegetarian combinations that may be deficient.] A purely vegetarian diet tends to decrease sexual drive in men since certain proteins and amino acids are in lesser quantities. Such a diet may tend to induce a more relaxed state, with decreased aggressive drives and a more contemplative attitude."[49]

This comment underscores the importance of obtaining amino acids in the proper balance.

HEAL FRACTURES FASTER

It happens suddenly. You slip and hurt your hip, arm, shoulder or any other body part or parts. And just as suddenly, you develop a fracture. This can be serious at any age, but more so in later years. Your chances for recovery are much greater if you have adequate amounts of amino acids as well as minerals such as calcium.

In one study reported by the Burke Rehabilitation Center of White Plains, New York, researchers noted that forty fracture patients, aged forty-three to eighty-six, were "starving" for more protein and calcium. These are essential bone-strengthening nutrients. The researchers noted that these fracture patients had a serious problem—they were losing very large amounts of nitrogen. This was a noticeable symptom that their bodies were speedily using up protein. A chronic loss could cause wasting away. "In the absence of aggressive therapeutic nutritional measures, total nitrogen and energy losses incurred in the thirty days following the traumatic episode may seriously prolong recovery of the patient," says the report.

Do you need more than the Recommended Dietary Allowance in this situation? The researchers say that on this minimal amount, you would need about fifty days to make up your nitrogen losses. You could cut it down to twenty days if you took twice the recommended amount. The report adds: "Patients recovering from fracture require far more protein than the approximately two ounces per day recommended for healthy adults; and the more severe the injury, the greater the need."

Additional calories are also needed because a fracture patient's energy needs are from 10 to 50 percent above normal. If your body is denied sufficient calories, it burns protein seized from the recovery process. The extra calories from carbohydrates and fats are called "protein-sparing" because they assist even minimal amounts of amino acids in the healing process.

Calcium is another nutrient needed by fracture patients. Calcium is a major constituent of bone, and researchers at the Burke Rehabilitation Center note that "repair of fractures would be hindered in the face of an inadequate supply of dietary calcium." They recommend that "calcium intake should be increased either by diet or supplements not only in the immediate postfracture period but also maintained throughout life to minimize future fracture risk." They suggest 1000 milligrams of calcium every single day, particularly for older people.[50]

IMPROVE WOUND HEALING

The slightest wound opens the way to invading hordes of bacteria; your body needs stronger tissues and blood vessels to block this threat. It can do this with amino acids, via protein, which act as building materials for the repair task of closing and healing wounds.

"Even more or less minor wounds require a good nutritional state and normal protein metabolism for optimal wound healing to take place," advises Sheldon V. Pollack, M.D., chief of dermatologic surgery at Duke University Medical Center.

Protein deficiency can be risky. It can slow down reconstruction of tissues and reduce your body's ability to protect itself from infection, according to Dr. Pollack. When recovering from an injury, boost your intake of protein foods such as fish, milk, eggs, cheese, liver and wheat germ.[51]

PROTECT AGAINST TOXICITY

As a mineral, copper is helpful in blood enrichment, among other benefits. But you may have elevated copper levels, perhaps from soft water carried through copper piping. This could cause an accumulation of copper in the brain and liver, leading to schizophrenia, among other problems.

Researchers have found that amino acids, in adequate amounts, are able to neutralize the potential toxicity of copper overload. Specifically, the sulfur-containing amino acids (cysteine, histidine, methionine) act as chelating substances; that is, they seize hold of these excess metals and promote their removal from the body. They also are said to reduce the toxicity of excess cobalt and molybdenum, and protect the body against excesses.[52]

UNTOASTED BREAD BETTER AMINO ACID SOURCE

Should you toast bread? Not if you want to retain the amino acids. As the bread darkens, the protein potency is reduced. You lose from 5 to 10 percent of lysine and from 15 to 20 percent of thiamine. And you also lose appreciable amounts of arginine, histidine, methionine, tryptophan and tyrosine. Thin slices have greater thiamine losses than thick slices because of greater heat penetration. There is also the risk of mutagenic (cancer-causing) compounds in excessively charred or browned bread.

Want toast anyway? Then have your bread just *lightly* toasted.

HEAL FASTER WITH AMINO ACIDS

Postoperative patients given amino acids and vitamins tend to heal better, according to Gerald Moss, M.D., Ph.D., of the Rensselaer Polytechnic Institute. In a talk before the American College of Surgeons, he told of giving ninety-four postoperative patients a combination of amino acids and vitamins as well as intravenous fluids and electrolytes. Their rapid improvement was attributed to this nutritive intake.[53]

FOOD FOR YOUR BRAIN

Can you think better with amino acids? Apparently so, according to Jose A. Yaryura-Tobias, M.D., medical director of Bio-Behavioral Psychiatry in Great Neck, New York. This leading psychiatrist believes that the amino acid tryptophan, together with the B-complex vitamins niacinamide and pyridoxine, increases the blood level of serotonin, the substance that promotes nerve impulses and influences behavior.

"When a patient is diagnosed as an obsessive-compulsive, the nutrients that we use are tryptophan, niacinamide and vitamin B6 (pyridoxine)," says Dr. Yaryura-Tobias. He prescribes tryptophan to ease depression, but recommends higher doses of prescribed vitamins B1 and B6, "because they help activate the energy transport system in the body. We also use phenylalanine. That is an amino acid which in the body is converted to phenylethylamine, an antidepressant."

Psychiatrists at the Great Neck facility use drugs when they are a necessity, but always together with nutrients. They do not limit themselves to just one type of therapy. There are many causes of illness and therefore various methods of treatment. Dr. Yaryura-Tobias tells emotionally upset patients that natural therapies—for example, the use of amino acids and nutrients—take longer than drug therapy, which can also have side effects.

"With the tryptophan and vitamins, results will be gradual, taking maybe ten weeks to reduce symptoms completely. But the benefits here are obvious—no side effects to mess you up in other ways."

If these methods do not work completely, the doctor will use behavior therapy, most often in conjunction with a proper diet and supplements; these are usually vitamin B6, niacinamide and tryptophan.

"But I can tell you this," notes Dr. Yaryura-Tobias, "something we are doing here is good, because even with the very sickest patients, we get about 50 percent to improve. And it's not the medication. As a psychopharmacologist, I'm sure of that. It's the addition of the vitamins. I'm *positive*. We also see that behavior therapy is helpful and so, too, is working with the patient's family. That's our approach and it works. I think that an integrated practice is the medicine of the future."[54]

* * *

With all the above information in mind, we may consider protein as a horn of plenty, a grab bag that contains an assortment of healing helpers, the amino acids.

8 AMINO ACIDS FROM VEGETARIAN SOURCES

THERE IS full amino acid power in meatless foods, when taken in the proper combinations. Thus a complete protein may be ingested at a vegetarian meal, even though no one food contains such a protein. Let's see how this is possible.

THE PROTEIN EFFICIENCY RATIO (PER)

"Proteins may be classified according to their source, such as animal or plant; or in terms of their essential amino acid composition; or on the basis of their PER, an index to their biological value, high or low," explains *Eat to Live*, a publication released by the Wheat Industry Council. "In general, animal proteins contain amino acids in proportions approximating human needs. Thus a relatively small amount of animal protein will supply all the amino acids needed in the human diet. Animal protein may be described as having a high biological value, or a high PER.

"The PER is determined by carefully controlled animal feeding tests and represents the amount of weight gained in grams for each gram of protein consumed. Casein, a milk protein, is used as a standard or control in such tests.

"Most plant proteins contain one or more of the essential amino acids at a level too low to match human requirements. They are classified as proteins of 'low biological value,' or low PER. Example: Wheat contains all of the essential amino acids, but is low in lysine.

"Other plant proteins, such as those of legumes like beans and peas, have fairly high PER values, although not as high as animal proteins. Furthermore, not all plant proteins are low in the same essential amino acid. There are differences in the amino acid content of various cereals, such as wheat, rice, oats and rye, as well as in legumes.

"A diet that includes a variety of vegetable or plant protein will usually furnish adequate amounts of essential amino acids. Thus by careful planning it is possible to meet protein and amino acid needs following even a strict vegetarian diet.

"The PER value of plant proteins can be increased by adding needed amounts of proper amino acids. Example: The essential amino acid lysine is added to some wheat-based cereal products to improve the protein quality. In addition, inclusion of some animal protein in a diet based largely on cereals improves its overall protein value. Eating bread and milk together provides a total protein intake of fairly high PER."[55]

FOOD SELECTION FOR MEATLESS PROTEIN

A concise guideline is suggested by Malden C. Nesheim, Ph.D., nutrition professor at Cornell University, in *The Medicine Called Nutrition*: "A continuing normal level of protein synthesis requires the continuous availability of the entire array of necessary amino acids. Removal of even ONE essential amino acid leads rapidly to a lower level of protein synthesis. Vegetarians should eat a wide variety of grains, cereals, vegetables and fruits to get the essential amino acids. A well-balanced, nutritionally adequate vegetarian diet is possible when generous amounts of legumes and whole-grain products and some milk, cheese and eggs are included daily. If dairy products are omitted on a strictly vegetarian diet, intake of vitamin B12 will be low.[56]

A VEGETARIAN GUIDE FOR AMINO ACID BALANCE

Vegetarians can be classified according to the types of animal foods in their diets. We classify meatless eaters in three basic groups:

1. *Vegans or Pure Vegetarians*. They consume plant foods only—no animal products such as meats, poultry, seafood, eggs, dairy.
2. *Lacto-Vegetarians*. They eat plant as well as dairy products, but nothing from the meat group. That is, they will consume milk and cheese products but no eggs.
3. *Lacto-Ovo-Vegetarians*. They eat plant foods as well as dairy products and eggs.

If you are in the first group, you will need to follow suggestions on how to pair plant foods properly for amino acid balance. The second group will have better amino acid balance from dairy products. The third group will be getting high-quality protein from the dairy and egg products.

STRICT VEGANS

*(*Recipes are included—see Chapter 12.)*

"PAIR"	*FOOD*
Legumes plus grains	Black beans and rice
	*Kidney Bean Tacos
	*Tofu, Rice and Greens

"PAIR"	*FOOD*
Legumes plus seeds	*Split Pea Soup with *Sesame Crackers
	*Garbanzo and Sesame Seed Spread
Legumes plus nuts	*Peanut and Sunflower Seed Tacos
	Dry-roasted soy beans and almonds
	*Chili Garbanzos and mixed nuts

LACTO-VEGETARIANS

"PAIR"	*FOOD*
Grains plus milk	Oatmeal and milk
	Macaroni and cheese
	Bulgur wheat and yogurt
Legumes plus seeds plus milk	Garbanzo beans and sesame seeds in cheese sauce
Legumes plus nuts plus milk	Mixed beans and slivered almonds with yogurt dressing
Legumes plus milk	Lentil soup made with milk
	Peanuts and cheese cubes
Seeds *OR* nuts plus milk	Sesame seeds mixed with cottage cheese
	Chopped walnuts rolled in semi-hard cheese

LACTO-OVO-VEGETARIANS

"PAIR"	*FOOD*
Legumes plus egg	Cooked blackeye peas with egg salad
	Buckwheat (kasha) made with egg
Grains plus egg	*Potato Kugel
Grains plus egg plus milk	*Rice and Raisin Custard
	*Cheese Muffins
	Cheese omelette with sesame seeds
Seeds plus egg plus milk	

MEATLESS EATING CAN BE BENEFICIAL

Since meat does contain saturated fats that are not desirable in excess, there is much to be said for a vegetarian program. These benefits are described in *FDA Consumer*:

"The high content of fat in the U.S. diet contributes to obesity. Overweight people are more prone to heart disease, high blood pressure and diabetes. One fills up fast on grains and such. Consequently, fat vegetarians are a rare species.

"If certain combinations of plant foods are eaten or if plant foods are taken with some animal protein, the missing amino acids are filled in. A number of these combinations have been around a long time, indicating that some of those old wives who told tales knew what they were talking about." Examples include:

Baked beans and brown bread, with one providing the amino acid the other is low in.
The same is true for macaroni and cheese.
Try a bean and cheese combination served as tamale pie.
Lentil–brown rice soup makes a good protein combination.
Wholegrain breakfast cereal and milk.
Dairy products, including eggs, provide high-quality protein and are good meat substitutes.
Legumes (soybeans, chickpeas) and nuts, including peanuts and meat substitutes made from plant protein (usually with a soybean base) are also protein sources for vegetarian diets.

Vegetarians are advised to provide themselves with two servings daily of the high-protein meat alternatives of legumes, nuts, peanuts, meat analogs and dairy products.[58]

"It is possible to be a vegetarian and enjoy a nutritionally sound diet with careful planning," comments Dr. Johanna Dwyer, director of the Frances Stern Nutrition Center at Tufts Medical School in Boston. "And there may indeed be some positive health benefits to such regimens. However, owing to lack of planning, a small group of vegetarians consume diets that are nutritionally incomplete and that jeopardize their nutritional status. The quality of vegetable proteins in the diet is improved by combining those legumes having a high concentration of certain amino acids with grains providing complementary amino acids. For example, cereal grains, which are low in the essential amino acid lysine but adequate in methionine, are complemented by legumes having adequate lysine but too little methionine. The overall value of the protein mixture is even better if a small amount of higher quality animal protein, such as milk, is included. With careful planning, even diets solely of plant foods can be satisfactory with respect to protein. Mono-diets, such as those based on only one cereal grain, should be avoided, especially in feeding young children."[59]

There are some studies to indicate that a meatless diet is able to protect against cancer and atherosclerosis and related circulatory diseases. Dr. Dwyer tells us that pure vegetarians who consume no animal food—compared to non-vegetarians and to those vegetarians who eat eggs and/or dairy products—generally exhibit the lowest serum cholesterol levels. They have lower low-density lipoprotein cholesterol (LDL) coupled with higher high-density lipoprotein cholesterol (HDL) levels. And they show lower levels of triglycerides. Furthermore, vegetarians show lower incidences of diabetes mellitus, high blood pressure, obesity and smoking, to name just a few problems. Dr. Dwyer emphasizes the importance of dietary planning as a means of maximizing the benefits of a meatless program.

THE BIRCHER-BENNER MEATLESS PROTEIN PROGRAM

The world-famous Swiss health spa, Bircher-Benner Clinic, has long been known for promoting effective healing on a meatless program. The head of this clinic, Ruth Kunz-Bircher, author of *Eating Your Way to Health*, maintains that vegetable protein has greater benefits than animal protein:

"Protein combinations of purely vegetable protein or mixed vegetable and other protein have consistently higher value than meat, milk and egg protein. Furthermore, vegetable protein has proved to be of much higher value in combination with a diet containing regular quantities of uncooked fruit and vegetables than in combination with the same quantity of cooked food. Protein requirements as low as 35 and even 25 grams per day (instead of 70 grams) have been proven capable of sustaining full vitality, stamina, health and growth."[60]

What about animal protein? "The apparent therapeutic value of animal protein is in reality due only to a temporary stimulation, and the effect does not last. This is indeed a burden on the metabolism. If an especially high-protein diet is indicated, it is successfully provided by milk protein in fermented form, such as yogurt or cream cheese added to the protein contained in green vegetables, wholegrain cereals and nuts."

The clinic does not favor excessive amounts of protein. "We do not give any protein in diseases of the kidneys and certain heart complaints, not even foods rich in vegetable protein; and we do not suggest any animal protein (eggs, meat, milk and cheese) in allergic and skin complaints or in cases of high blood pressure.

"Supply of the most complete proteins of high biological value is possible by combining wholegrain cereals with green vegetables or milk protein. Soya beans, nuts, cereal germs and germinated pulses are rich in equally valuable vegetable proteins." (Germinated pulses refer to sprouts of beans, legumes and seeds, and are considered of top quality in amino acids.)

EASY WAYS TO ENJOY MEATLESS PROTEIN

Myron Winick, M.D. of the Institute of Human Nutrition at New York's Columbia University College of Physicians and Surgeons, notes that it can be easy to be protein-nourished on a meatless diet if you "complement" in meal planning.

"By simply combining different plant foods containing protein at the same meal, all the essential amino acids are present at one time. The resulting combined proteins are then complete. This practice is called 'complementing proteins' and is based on the fact that by eating two incomplete proteins in which different amino acids are missing, you end up with a complete protein."[61]

Dr. Winick suggests these tasty meatless combinations:

Chili *sin* carne (*without* meat) with rice
Baked beans and brown bread
Okra, corn, lima and tomato gumbo
Macaroni and cheese
Zucchini or spinach quiche
Bean sprouts and tofu with rice
Squash stuffed with cottage cheese or tofu

PROTEIN SOURCES AT-A-GLANCE

FOOD	*AMOUNT*	*PROTEIN (Grams)*
Rice (1 cup) and beans (½ cup)		10
All milk and yogurt	8 ounces	8
Cheese (sandwich type)	1 ounce	7

FOOD	AMOUNT	PROTEIN (Grams)
Cottage cheese (pot style or regular)	½ cup	14
Tofu (2″ × 3″)	120 grams	9.4
Leafy green vegetables	1 cup	4 to 5
Most beans (cooked)	1 cup	14 to 15
Peanut butter	2 tablespoons	8
Most nuts	½ cup	7 to 12
Bread	1 slice	2 to 3
Noodles (cooked)	1 cup	7
Brewer's yeast	1 tablespoon	3

Source: USDA

Whether you want to eat less animal food or no animal food, you can have your complementary balance of amino acids with some simple planning ahead. And good taste will go along with good health rewards from a meatless protein way of life.

9 POWER PROTEIN FOODS AND SUPPLEMENTS

WHEN PLANNING your amino acid program, you have a wide variety of foods to help fill your nutritional needs. You have to select those foods that give you complete protein. If there is a missing essential amino acid or one low in needed proportions, then there is a limiting of the entire process of protein synthesis.

For example, you eat a cornmeal product; it is low in tryptophan. Protein manufacture keeps on in your cell as long as there is available tryptophan. When this amino acid is depleted, then construction of your cell closes down, regardless of the availability of other amino acids. That is why you need to select a variety of foods, especially if you are a vegetarian, to give you balanced amino acids that promote protein power in cellular building.

If any food's amino acid availability entirely matched your body's requirements, its level of protein usability would be 100 percent. "Biological value" is the term used to describe the percentage of any food's protein usability. An egg has a high biological value because more of its amino acids are used by the body than those in any other food except human milk.

UNDERSTANDING NET PROTEIN UTILIZATION OR NPU

"The blanket term used to describe both a food's biological value and its relative digestibility is *net protein utilization*, or NPU," explains Vic Sussman in *The Vegetarian Alternative*. "The egg, with its ease of digestibility and high biological value, has an NPU of 94 percent. That is, almost all of the egg's protein will be available to the body when the egg is digested. Other foods with either less digestibility or a less-ideal amino acid pattern will have correspondingly lower NPU ratings."[62]

What about animal foods? Author Sussman feels that meat and poultry, while traditionally reputed to be superior in terms of NPU, are really not. Fish, the exception, has an NPU of 80, coming close to eggs. In contrast, most meat and fowl rate only about 67.

As for dairy products, "milk seems low in protein (3½ to 5 percent) if you consider only quantity, but its NPU is an efficient 82 percent; skim, whole, buttermilk and yogurt all have the same rating. Most cheeses have an NPU range of 70 to 75 percent. You could stop eating flesh foods today and easily get plenty of protein from milk products and eggs," notes Sussman.

UNDERSTANDING THE NITROGEN CONVERSION FACTOR, OR NCF

Protein is composed of 16 percent nitrogen, an essential constituent. But your body constantly gives off nitrogen through wastes as well as through perspiration, skin, nails, and hair. You could run the risk of a nitrogen deficiency. Just a small intake of protein daily will give you adequate amounts of nitrogen, the substance you need to create the protein of blood, tissue and regulatory fluids. A deficiency could cause malfunctioning of these processes.

The Food and Nutrition Board suggests that an average person requires about 0.213 grams of protein per pound of body weight every single day. A 154-pound person needs about 33 grams of *usable* protein every day to maintain nitrogen balance. This process is referred to as the Nitrogen Conversion Factor or NCF. You need a minimum amount of daily protein to make nitrogen available for body rebuilding. Some foods offer a higher NCF than others. And there are times when you do need more nitrogen to meet the challenges of daily living.

Nutritionist Henrietta Fleck cautions that certain emotions can be thieves of nitrogen: fear, pain, and anxiety, anger. As much as one-third of your supply of nitrogen can be dissipated in stressful situation.[63]

"The cumulative minor stresses of life in a competitive society" also causes a nitrogen drain, says the Food and Nutrition Board. Exposure to extremes of heat or cold, excessive perspiration and non-disabling but nitrogen-draining infections all take their toll.[64] If you are all "tied up in knots," there could be a blockage of amino acid absorption in your bloodstream. Emotional and/or physical pressure can further impede nitrogen conversion so that you run the risk of a deficiency.

N.S. Scrimshaw in the *American Journal of Clinical Nutrition* explains that "the degree of nitrogen loss is dependent upon the severity of infection or injury; obligatory losses may approach the equivalent of 0.9 grams of protein per day."[65] Nutritionist B.R. Bistrian, in the *Journal of the American Medical Association*, says that "prolonged infection and injury without adequate nutritional support will lead to protein malnutrition, as has been observed in hospitalized adult patients."[66]

"The protein-nitrogen content of diet should be considered independent of calorie content," emphasizes George L. Blackburn, M.D. of Harvard Medical School in *Surgical Clinics of North America*. "Nitrogen:calorie ratios are not instructive in designing dietary therapy. At a caloric intake of 45 calories a day per pound of body weight the desirable protein intake would be 1.8 grams to allow for optimal nitrogen utilization. At these levels, the ratio of nitrogen to calories is approximately 1:150. Special considerations for patients with liver or renal disease must be taken into account in terms of protein requirements. During acute phase of injury, nitrogen intake may need to be further increased, without an increase in non-protein calories."[67]

A reasonable margin of amino acids over the minimum does make good health sense. You have a variety of animal and plant foods to select from that have good degrees of Net Protein Utilization (NPU) or Nitrogen Conversion Factor (NCF) to give you the edge on better health and protection against illness. The chart appearing at the end of this book gives you an assortment of everyday foods with varying degrees of NPU-NCF potencies. Additionally, there are a wide variety of foods that can be included in your quest for amino acid improvement in your daily diet. These include:

ACIDOPHILUS. A soured milk preparation, fermented by the natural therapeutic bacteria *lactobacillus acidophilus*, needed to process food that might otherwise putrefy. A good source of amino acids, as potent as the milk from which it is made.

ALFALFA. Whether as a seed, leaf or sprout, a very good source of plant protein and many essential amino acids. Plan to use alfalfa as often as possible for its nutritional treasures.

ALMONDS contain 84 grams of high-quality protein per pound. Best eaten raw, in their natural state.

BIRCHERMUESLI. A popular Swiss-created breakfast food with a rich source of amino acids via its content of nuts, seeds, whole grains. Mix with milk for better complementary balance; sprinkle brewer's yeast on top for more amino acid potency.

BRAN. The broken coat of the seed of cereal grain separated from the flour or meal by sifting or bolting. A prime source of amino acids. Sprinkle over cereals, use in baked goods and in blenderized beverages, hot or cold.

BREWER'S YEAST. At least 50 percent protein, it has an NPU of 67. Mix with beverages, soups, blender recipes, baked recipes, bread doughs, casseroles.

BUCKWHEAT. The triangular seed of the cereal grass that offers concentrated protein and many essential amino acids. Combine with other wholegrain flours for better balance.

CASHEWS. A kidney-shaped nut of a tropical American tree of the sumac family, it has a high protein content, and is high in unsaturated fats. Cashew nut milk may be used to replace whole dairy milk, if desired.

DESICCATED LIVER. Dried, powdered beef liver, with the fat and connective tissue removed. A source of complete protein. Stir in a glass of vegetable juice; add to soups or stews. While not tasty, it can be flavored with a squeeze of lemon or lime juice.

DULSE. Dried red seaweed used in Scotland, Ireland and other northern countries as food; it has about 25% protein, depending upon variety.

FARINA. A fine meal of assorted grains that boosts the power of amino acids; an easily digested breakfast cereal or pudding.

FISH LIVER OIL. A good source of protein, offers almost a complete balance of amino acids.

GARBANZOS. Also known as chickpeas or cici beans; an extremely high source of vegetable protein. Combine with sesame seeds to make a complete amino acid food.

GRAINS. These are small, hard seeds or fruits or edible cereal grasses. Unrefined, they provide potent amino acids that need to be complemented with dairy foods or egg for complete balance. Varieties include wheat, millet, corn, brown rice, oats, buckwheat, rye, triticale. Use as flour for baking or in cereal form. Available as ready-to-eat mixtures of grains, seeds and nuts for complete protein.

GRANOLA. A high-protein cereal mixture. Different varieties are available made up of wheat, wheat germ, oats, sesame seeds, honey, dried fruits, bran, rolled oats, nuts, dates, coconut flakes, vanilla. Add to dairy milk or soya milk for complete protein.

LECITHIN. A bland, water-soluble granular substance usually derived from defatted soybeans or from sunflower seeds. Available as tablets or in granular form; add to wholegrain cereals or baked goods and aim for complementary protein boosting.

LENTIL. A legume, a prime source of many amino acids along with vitamins, minerals, iron.

MACARONI. A pasta, made of flour or artichokes, and a prime source of many amino acids. When cheese is added, the complementary pattern is complete.

MILLET. Oldest of grains in human use. Has most of the essential and non-essential amino acids; its protein content is said to equal that of animal protein. Available hulled, it is a delicious cooked cereal.

NUTS. A prime source of meatless protein. Enjoy unsalted nuts for good taste, as nut butters, or ground and added to baked goods.

PASTA. Made with soy, artichoke, wholewheat or buckwheat flours to come close to complete protein. available in many shapes and sizes.

PEANUTS. High in protein, a good source of B-complex vitamins. Can also be blended to make milk for cooking purposes.

POLLEN, BEE. The male sexual substance of seed-bearing plants. A prime amino acid food that is said to contain all of the twenty-two elements of which the human body is made.

PROTEIN SUPPLEMENTS. A great range of products in capsules, tablets, wafers, powders; some contain all essential amino acids along with other helpful ingredients. These are protein-rich cereals, protein "chews" for a snack, liquid protein supplement drinks and instant protein soup mixes. Read labels carefully to determine the amino acid content and potency for your specific needs.

PUMPKIN SEEDS. Nearly 60 on the NPU scale, one tablespoon will provide a bit more than two grams of usable protein. Very high in such essential amino acids as isoleucine and lysine; combine with grains low in lysine for a good complementary protein balance.

RICE, BROWN. While modest in protein, brown rice has a high NPU of close to 70.

SEAWEED. Various sea plants are available with varying levels of amino acid content.

SESAME SEEDS. With an NPU of 53, they can be a good source of many other nutrients such as calcium, B-complex vitamins (especially niacin, choline, inositol) and also vitamin E.

SEVEN-GRAIN CEREAL. Combination of seven unrefined grinds of wheat, corn, barley, oats, rye, soybeans, bran, rice bran to be used as cereal or added to casseroles, puddings, loaves, etc. You can also replace a small amount of flour with cooked cereal when baking. Very high in protein content, rivaling that of meat.

SOYBEANS. The amino acid balance is believed to be similar to that found in meats. It has an NPU of about 56 to 61, depending upon whether you use flour (61), granules and beans (57) or sprouts (56). You can include soybeans in these forms or as tofu and tempeh, among other products.

SPROUTS. Germinated seeds. Of any sort, they are prime sources of high levels of NPU and NCF, since the sprouting process increases the amino acid and nutrient content to mega-proportions. Sprouts can be grown right in your kitchen; if purchased, make certain they are free of sprays or chemicals. Enjoy them raw as part of a salad, in a sandwich, soup, etc.

SUNFLOWER SEEDS. One tablespoon supplies a bit more than one usable gram of protein.

TAHINI. Made from sesame seeds, it may offer as much as 50 percent protein, along with essential vitamins and minerals.

TVP or MEAT ANALOGS. Textured vegetable protein is made from a meatless source that looks and tastes just like the animal food. You can enjoy chicken, turkey, ham, pastrami, steaks and burgers, but without any meat. These analogs are made from soybeans, peanuts, wheat, and other vegetable products and spun into an imitation meat product. Many TVP products contain added vitamins; some have eggs, so this adds cholesterol. Caution: check labels for artificial colors, additives, salt, monosodium glutamate (MSG) or other additives which should be avoided.

WHEAT GERM. Its protein utilization or NPU is 67, which puts it on a level comparable to animal products. One tablespoon of wheat germ will provide about one usable gram of protein. It is low in tryptophan, so complement with a milk product.

YOGURT (unflavored) will offer the same NPU as milk (along with the same number of minerals and calories), but will offer better amino acid assimilation. The reason is that the bacteria that create fermentation also predigest the lactose and this leads to better utilization.

At a Glance. Milk has an NPU of 82. Egg boasts an NPU of 94. Both will provide more usable amino acids than would animal foods such as meat, poultry or fish. Cheese gives you an NPU range from 70 to 75. Mix milk or eggs with some plant protein, and you boost the total NPU of the meal. A small amount can have a powerful effect in terms of fortifying your body with all the needed amino acids. *Suggestion*: Milk is a great source of amino acids. Mix two tablespoons of powdered milk with a cup of rye or wheat flour when you bake. You will boost the overall NPU and NCF values by at least 50 percent.

Put power into your body with amino acids from a wide variety of protein foods and supplements. You'll feel the rewards in no time at all!

10 FOR WOMEN ONLY: Amino Acids Are Potential Lifesavers

PROTEIN and amino acid requirements vary according to age, weight and sex. Because women go through three physiological processes not experienced by men—PMS or premenstrual syndrome, pregnancy-lactation, menopause—they need more amino acids in the proper balance than men do. The critical times of a woman's life can be helped with the use of amino acids; they are perhaps, more important during these three periods than at any other time.

PMS OR PREMENSTRUAL SYNDROME

About 76 percent of women between the ages of fifteen and forty-four suffer from menstrual cramps and up to 90 percent of all women in their childbearing years experience at least some symptoms of the premenstrual syndrome, which may go beyond emotional manifestations such as irritability and depression, and include bloating and cramping, among other symptoms. This would put the figure at over 25 million women who have reactions about ten days out of every month.[68]

It has been suggested that the mood swings, weight gain, and violent reactions of PMS are caused by hypoglycemia, or low blood sugar, during the period when hormones are in constant flux. To help stabilize the situation, a protein-amino acid program is often advised to ease the symptoms of hypoglycemia.

Hypoglycemia is defined as a deficiency of glucose or sugar in the bloodstream, causing muscular weakness and lack of coordination, mental confusion, sweating and nervous reactions. Treatment should be directed at correcting nutrition, notes leading New York gynecologist Niels H. Lauersen, M.D., co-author of *Premenstrual Syndrome and You*. A woman who has food cravings and hypoglycemic symptoms "should be sure to eat at approximately three-hour intervals in order to keep her blood sugar steady," says Dr. Lauersen. "She should avoid foods with refined carbohydrates and eat meals and snacks that include protein and natural carbohydrates. A woman may choose from cheese, yogurt, nuts, sunflower seeds, eggs, her favorite fruits and vegetables, and unrefined wholegrain products, which contain more nutrients than sugary sweets."[69]

Dr. Lauersen says that if you consume carbohydrates, have protein at the same time, but *never* have a carbohydrate without the protein.

He explains why: "Carbohydrates enter the bloodstream rapidly and create a demand for insulin to be used in glucose conversion. The insulin climb then may cause a drop in blood sugar, which in turn triggers a PMS mood swing. If, for example, an apple or an orange is eaten with a piece of cheese, which is carbohydrate paired with a protein, the cheese will temper the effect of the fruit, insulin production will be less dramatic, and the blood sugar will not descend to the symptom-causing level."

When the PMS victim is struggling to keep her blood sugar from undergoing drastic fluctuations that produce eating urges, "the carbohydrate/protein combination, or protein alone, at regular intervals" should bring the desire under control. Dr. Lauersen adds, "A woman should choose natural carbohydrates and protein foods such as fish and chicken, which are lower in fat and calories than red and processed meats. A hard-boiled egg is high in protein and contains only 75 calories."

Patricia Allen, M.D., co-author of *Cycles*, agrees that a carbohydrate/protein combination can help ease PMS distress. This New York gynecologist suggests that "the most advisable eating strategy during the premenstrual week involves frequent smaller meals consisting of high protein and complex carbohydrates, with no refined sugars or starches and a minimum of salt and exotic spices.

"For protein, good alternatives to red meats include grains, seafood, poultry, low-fat dairy products (also excellent for calcium) and legumes (in moderation). Another source, desiccated liver, has the advantage of being cholesterol-free."[70]

Earlier in the book, Dr. Allen points out that "Premenstrual headaches are sometimes thought to be the result of metabolic disturbances such as blood sugar fluctuations or sensitivities to certain foods. When the blood sugar is relatively low, the body often compensates for the deficiency by sending a greater volume of blood to the brain. The result of this higher pressure and 'stretching' of blood vessels could be throbbing head pain."

To take the sting out of cramps, an easy program is suggested by Leo Wollman, M.D., a Brooklyn, New York-based physician and author of *Eating Your Way to Better Sex Life: The Complete Guide to Sexual Nutrition*. "During the premenstrual week, it will be helpful to put yourself on a hypoglycemia diet of several high-protein meals in smaller proportions. Red meats, which raise your serum acidity, should be avoided at this time. Choose fish, poultry, cheeses and vegetable proteins. Avoid coffee and sugar. Caffeine is known to exacerbate dysmenorrhea [painful or difficult menstruation]."[71]

PREGNANCY AND LACTATION

The Recommended Dietary Allowances for amino acids/protein increase during childbearing and lactation.

The importance of this nutritional boosting is described by C. W. Whitmoyer, Sr., D.Sc., author of *Your Health Is What You Make It*. He says: "Pregnancy and lactation are circumstances which make it urgent to increase protein intake. Under these circumstances, a woman should certainly increase her protein intake by 20 percent over the amount considered adequate prior to pregnancy. This assumes that her nutrition was at a satisfactory level for a prolonged period prior to her pregnancy.

"Many young mothers (more than one-third of births are to women under twenty-one years) have followed a poor diet consisting too largely of such items as potato chips, candy, soft drinks and doughnuts. These women are in a very poor state of nutrition to nourish a developing baby and need to increase their protein intake sharply if they are interested in protecting the welfare of their own bodies and properly nourishing their fetuses so that they may deliver normal, healthy babies.

"A great deal of protein is required to form the multiplicity of types of tissues that are necessary for the development of the complete little body growing daily in the womb. There is no source of this required protein other than the mother's own tissues and the food that she consumes. Likewise, if the baby is breast-fed after birth, the mother will have a heavy output of protein, since human milk is a high-protein product, containing an abundance of essential amino acids. There is no source of these milk proteins other than the mother's own tissues and the food she consumes. There are no circumstances in which protein nutrition becomes a more critical matter than during the period of pregnancy and lactation."[72]

Some easy ways of increasing amino acid intake during pregnancy are suggested by nutrition authority Vic Sussman in *The Vegetarian Alternative*. He explains that the Recommended Dietary Allowance (RDA) for expectant women would call for 30 grams a day more protein than normal, or about 76 grams daily. This can be accomplished in the following ways:

- Increase use of milk and eggs. Skim milk powder is low in fat, easy to add to almost any dish or liquid. Non-instant milk powder contains about 8 grams of usable protein per one-quarter cup.
- A few ounces of cheese can go a long way in boosting your amino acid intake during pregnancy.
- One-half cup of uncreamed cottage cheese gives you about 25 grams of usable protein.
- Add full-fat soy powder (not flour) to blender drinks, cereals, breads, soups, casseroles.
- Soy flour boosts amino content of breads and muffins.
- Grains, legumes and nuts can be used for fortification.
- Remember to include brewer's yeast, which has 50 percent usable protein and a storehouse of vitamins and minerals.
- Use tofu or soybean curd, giving you 65 percent of usable and highly digestible amino acids.

Sussman adds, "Soy powder, high-protein bread, grain-legume and seed-grain dishes, brewer's yeast and tofu can help expectant vegans to make up the difference comfortably."

Finally, he suggests, "You may want to add up your daily intake of protein for several days to determine your range of intake; but don't stuff or starve yourself according to a chart and a column of numbers. Eat a sensible diet drawn from a variety of sources."[73]

MENOPAUSE

Taken from the Greek words *men*, meaning "month" and *pausis*, meaning "stop," it refers to that time of life when a woman's monthly menstrual period comes to a halt. There may be such familiar change-of-life symptoms as hot flashes, backaches and dry skin, to name a few. Because of changes in the hormonal balance, there may also be emotional upset, depression and weight gain. The goal here is to help provide amino acids in sufficient amounts to help lessen the severity of such symptoms. Because amino acids enter into the manufacture and distribution of the body's hormones, these nutrients do play an important role during this time of life.

One problem associated with menopause is that of osteoporosis or a thinning of the bones. This could lead to fractures and disabilities, even fatalities because of a loss of mass from the bone structure. To guard against such a hazard, women should increase their intake of calcium and phosphorus, two minerals that work with protein to help build strong bones. The anatomy of the bone is described by Morris Notelovitz, M.D., author of *Stand Tall!* and director of the Center for Climacteric Studies at the University of Florida in Gainesville.

"Bone tissue is composed of tiny crystals of calcium and phosphorus embedded in a framework of interlocking protein fibers. These protein fibers are made primarily of collagen. It is the calcium crystals that give your bones their strength, hardness and rigidity, and the collagen fibers that give them their relative capacity for flexibility.

"A number of other materials are also present in bone—fluoride, sodium, potassium, magnesium, citrate—as well as a host of trace elements. These act as the 'mortar' holding the 'bricks' of calcium and phosphorus crystals together."[74]

Dr. Notelovitz calls for supplementation of calcium to amounts as high as 1000 to 1500 milligrams a day. But he makes it clear that calcium can be lost by a protein imbalance. Therefore, the goal is to have more calcium but less protein. Higher biological values of protein should be obtained from grain and meatless sources that are less likely to cause loss of calcium. With this simple adjustment, you can help build a strong bone structure and protect against osteoporosis.

According to Dr. Patricia Allen, "It has been suggested that red meats, caffeine, excessive alcohol, refined white sugar and chocolate can all aggravate menopausal symptoms and perhaps should be avoided or kept to a minimum. A diet plentiful in fresh vegetables and fruits, grains, seafood, legumes and low-fat dairy products is advisable."[75]

Another problem in menopause may be blood sugar imbalance. Correct blood sugar upset and you may be able to correct your symptoms. The same high-protein program that sends amino acids through the body to stabilize blood sugar and correct hypoglycemia may well ease distressing changes. According to Robert C. Atkins, M.D., author of *Dr. Atkins' Nutrition Break-*

through, "Hypoglycemia and menopause share a myriad of symptoms, many of which are amenable to the same dietary treatment. In particular, the depression and the mood swings of both conditions respond to blood sugar control. Those who have a difficult menopause may be having problems because they have low blood sugar in addition to 'change of life.' In fact, symptoms of low blood sugar may be accentuated in the menopausal patient just as they are in the patient with premenstrual tension."[76]

To stabilize blood sugar during menopause, remember the two principles of Net Protein Utilization (NPU) and Nitrogen Conversion Factor (NCF). You may be consuming protein foods that have low utilization and conversion reactions, and the subsequent deficiency could trigger off menopausal symptoms similar to those of low blood sugar or hypoglycemia.

The chart of amino acid foods at the end of this book shows you the higher NPU and NCF factors. Select from these foods and breeze through menopause.

11 SUPER FITNESS WITH AMINO ACIDS

A COMBINATION of exercise and amino acids can accelerate your circulatory-metabolic processes and create regeneration from within so that you radiate good health and fitness. Since much of what you see (and cannot see) consists of amino acids, you can appreciate the importance of nourishing your body with these nutrients and having them be metabolized properly. This depends in part upon regular exercise programs, easily performed and very effective in cell-tissue rejuvenation. Remember, there is life in movement! Wake up your sluggish metabolism with simple fitness programs. Your amino acids can then be transformed into building blocks of health.

There are four types of exercise, according to authorities in physical fitness, including the President's Council on Physical Fitness and Sports.

1. *Aerobic* exercises are the more vigorous types of activity designed to improve the organs and systems that help the body to process oxygen—the heart, lungs and blood vessels. These exercises, which include running, jogging, bicycle riding and swimming, help your lungs to process more air with less effort. This in turn strengthens your heart and increases your endur-

ance capacity. The key to aerobic exercise is continuous activity without rest periods. It is often referred to as long slow distance (LSD) activity.

2. *Anaerobic* exercise includes activity of short, intense duration followed by a period of recovery. Such activities as tennis, handball, and sprinting are examples of anaerobic exercise. These activities place a sudden, high demand on the heart and lungs.

3. *Isometric* exercises are strength-building activities that involve no actual movement. This kind of exercise is accompanied by pitting muscle groups against one another or against unyielding objects. This is usually done for a period of 10 to 15 seconds at maximum effort. Because of the limited movement, isometrics provide little functional strength development and in some cases, may limit joint range of motion. Furthermore, many doctors believe that elevation in thoracic (chest) pressure may cause dizziness and fainting in some individuals. This activity is not recommended by the President's Council on Physical Fitness and Sports.

4. *Isotonic* exercises are strength-building exercises that require using muscles through a full range of motion. Such activities as weight-lifting, push-ups, sit-ups and pull-ups are all examples of isotonics. This kind of exercise is important in maintaining muscle mass and sound posture.[77]

Weight is the net result of calories taken in compared with those used up in exercise or in maintaining your body and its functions. The calories taken in can be reduced only so far; many people feel uncomfortable on diets below 1200 calories. Exercise, however, is a variable that can be increased greatly, so try to choose a program that you will enjoy in order to avoid boredom. Vigorous exercise allows weight loss while allowing you to eat a satisfying diet; it also helps maintain muscle mass.

When you consume protein, it needs to be transformed into cell tissue and muscle-building amino acids. You need to maintain muscle mass and a firm figure. Amino acids help do this, but only if they are invigorated through activities such as exercise. Otherwise, there is atrophy of muscles from disuse; and this is not only a prime cause of aging, but it is also a key factor in creeping obesity. True, you may consume sufficient amino acids for your particular needs, but they remain inert and lifeless without activation from exercise.

The simplest and most effective exercise is *walking*! It is a highly desirable form of exercise because it requires no special training or equipment. You can achieve an adequate level of fitness by walking twenty miles a week as briskly as possible.

Begin gradually at a pace that is comfortable for you and work up to this goal. You can work it into your daily routine by avoiding elevators and climbing stairs when possible, walking to the local stores instead of driving and taking a daily brisk walk after lunch.

Good quality amino acids enter into the construction of increasing muscle mass, which takes place as a result of exercise. It is activity that breaks down protein into nitrogen-bearing amino acids to be absorbed by the blood and carried to the tissues. Here they are selectively removed by cells in need of specific amino acids.

During the process of life, many millions of cells are destroyed daily. If the amino acid supply is not sufficient to offset the daily destruction of cells, the body will begin to waste away. And remember that even if you consume an adequate amino acid pattern, you have to activate these substances to do their cell-tissue rebuilding and that is possible solely with fitness!

Select an exercise program that is comfortable, easy and effective. Stick to it. You will soon feel the rewards.

BEFORE YOU BEGIN

It is important to get off to a good start. If you are thirty years of age or older, a visit to your family doctor is suggested prior to beginning your program. This is especially important if you have been inactive for a period of time, are over age thirty-five or more than ten pounds overweight or have any medical problem.

Warm Up First

Before starting any exercise, sport or strenuous activity, you need to warm up. A simple warm-up is essential to elevate your heart and respiration rates and increase your circulation, vessel dilation and body temperature. Simple stretching helps ease stress and tension in your muscles, making them more pliable. You will then be able to do more with less chance of injury.

Give your body a chance to limber up within a five- to ten-minute time slot. Start at a medium pace and gradually increase it. Listed are three warm-up exercises. Each one stretches a different part of your body. Do these stretching exercises slowly and in a steady, rhythmical manner. You'll thus be preparing your joints and muscles for the selected exercise program.

1. Wall Push: Stand about 1½ feet away from the wall. Then lean forward pushing against the wall, keeping heels flat. Count to 10 then rest. Repeat one or two times.

2. Palm Touch: Stand with your knees slightly bent. Now bend from your waist and try to touch your palms

to the floor. Do not bounce. Count to 10 then rest. Repeat one or two times. *Note*: If you have lower back problems, do this exercise with your legs crossed.

3. *Toe Touch:* Place your right leg level on a stair, chair or other object. Keeping your other leg straight, lean forward and slowly try to touch your right toe—with right hand 10 times, with left hand 10 times. Do not bounce. Then switch legs and repeat with each hand. Repeat entire exercise one or two times.

During the warm-up, perform each stretch slowly and smoothly. Do NOT bounce, as this can tear or strain your muscles. Breathe deeply while you perform each movement and as you hold each stretch. Never stretch to the point of feeling pain. Alternate warm-ups could include rhythmic activities, jogging in place, arm circles and hip rotations.

Whatever your selected exercise, you need to keep track of your heart rate. Your maximum heart rate is the fastest your heart can beat. Exercise above 75 percent of the maximum heart rate may be too strenuous unless you are in excellent physical condition. Exercise below 60 percent gives your heart and lungs too little conditioning. The 60–75 percent range is called your target zone.

Aim for the lower part of your target zone (60 percent) during the first few months. As you get into better shape, gradually build up to the higher part of your target zone (75 percent). After six months or more of regular exercise, you can work out at up to 85 percent of your maximum heart rate—if you wish. However, you do *not* have to exercise that hard to activate your amino acids and maintain fitness.

Intensity of the selected exercise should be moderate. It can be monitored by maintaining your heart rate within a "training heart rate" range. To find your target zone, look for the age category closest to your age and read the line across.

AGE	*TARGET ZONE (60–75%)*	*AVERAGE MAXIMUM HEART RATE*
20 years	120–150 beats per minute	200
25 years	117–146 beats per minute	195
30 years	114–142 beats per minute	190
35 years	111–138 beats per minute	185
40 years	108–135 beats per minute	180
45 years	105–131 beats per minute	175
50 years	102–127 beats per minute	170
55 years	99–123 beats per minute	165
60 years	96–120 beats per minute	160
65 years	93–116 beats per minute	155
70 years	90–113 beats per minute	150

Caution: A few high blood pressure medicines lower the maximum heart rate and thus the target zone rate. If you are taking high blood pressure medications, call your physician to find out if your exercise program needs to be adjusted.

HOW TO COUNT YOUR PULSE

The pulse count is nearly always the same as the number of heartbeats per minute (the heart rate). When you stop exercising, quickly place the tips of two or three fingers on your other hand, just below the base of your thumb on the inside of your wrist. Apply light pressure to feel the beat. (Do NOT use your thumb since it has a pulse of its own that can be confusing.) Count your pulse for 30 seconds and multiply by two. You should count your pulse *immediately* upon stopping your exercise because the rate changes very swiftly once you either slow down or stop.

If your pulse is *below* your target zone, exercise a little more vigorously the next time. If you're *above* your target zone, exercise a little easier. If it falls within your target zone, you're doing fine.

Once you're exercising within your target zone, you should check your pulse at least once each week during the first three months and periodically thereafter. If you have difficulty breathing, experience faintness or prolonged weakness during or after the exercise, you are doing it too hard. Simply cut back and check your pulse to see if you are still within your target zone.

As you begin to improve your fitness level and become "trained," you may increase the vigor of your exercise. But always be sure to keep your rate heart in the target zone for at least 20 minutes.

HOW TO SET YOUR SCHEDULE

Your goal should be a minimum of 20 to 30 minutes every single day. Build up your exercise program gradually over the weeks until you reach this goal.

You must stick to it. You will maximize the amino acid cell-tissue building activity if you keep at it. Your activity should be brisk, sustained and *regular*. Don't push yourself too hard to the point where it is no longer any fun. But stick to your program.

If you've eaten a protein meal, or any other type, hold off exercising for at least two hours. Better yet, plan to exercise before your meals as a means of releasing hormones that control your appetite. Eat 30 to 60 minutes *after* your exercise for boosted amino acid metabolism.

Cool-Down Last

After exercising, allow your body to experience a slow cool-down. An abrupt stopping of any exercise can cause dizziness. Your heart, lungs and muscles need a chance to adjust. To help yourself cool down, try mild stretching, deep breathing, walking while swinging your arms.

You can also cool down by changing to a less vigorous exercise, such as from jumping rope to walking. This allows your body to relax gradually. Swim more slowly or change to a more leisurely stroke. If you have been running, walking briskly, or jumping rope, repeat your stretching and limbering exercises to loosen up your muscles.

Basically, the cool-down period assists in returning your body slowly to normal function, returning blood from exercised muscles back to your heart and helping prevent muscle and joint soreness.

Which Exercise for You?

To decide, use either the target zone or pulse rate system, to see if the exercise is suitable for your particular body. Select any or all of these and alternate between one and the other. They are all designed to wake up your sluggish metabolism and distribute amino acids to your network of cells and tissues where rebuilding can be performed. And they are fun to do, too.

*Brisk walking is one of the best ways to invigorate your system. Do this as often and as much as is comfortably possible.

*Jogging is popular, along with cross-country skiing or rowing. You will burn up at least 800 to 1000 calories per hour for any of these. Be sure to get doctor clearance for your personal condition.

*Roller skating. That's right, the fun game for kids is a popular one for adults; it is less jolting to your leg muscles and joints.

*Swimming is easy and can be done indoors or outdoors, depending upon your circumstances and location. A bonus here is that you are able to metabolize additional calories if the water is cool.

*Bicycle riding is a great way to supercharge your body with oxygen-carrying amino acids. You can schedule half to one hour of cycling every day; you can also bike your way to do errands and other chores.

*Basketball is a vigorous activity that will pep up your circulation and help boost the metabolism of protein molecules into usable amino acids. Be sure to keep moving to keep your pulse rate up.

*Tennis can be fun but you need to maintain a constant tempo in order for your metabolism to be revved up to a sufficient level.

The exercise you pick, whether working out at home or outdoors, by yourself or with others, should be *fun*. It should be healthful in terms of a better amino-acid nourished body.

SAMPLE DIET FOR RUNNERS

Julian Whitaker, M.D., a member of the American Medical Joggers Association and director of a Panorama City, California, health center, leans toward meatless protein as a healthy diet for runners, and, presumably, other fitness buffs. In "How Much Protein Do Runners Need?" he offers the following diet for runners:

Breakfast: Have a large bowl of hot wholegrain cereal, ripe banana for sweetener and non-fat dry milk. Also try several pieces of high-fiber wholegrain toast without butter or margarine and some fresh fruit.

Snack: Several pieces of fresh fruit. Munch on wholegrain bread as well.

Lunch: Have a fresh green salad with lemon juice, honey and spice dressing—no oil. Include a bowl of brown rice, wholegrain bread and fresh fruit.

Snack: Same as the morning snack.

Dinner: Eat a large dish of eggless pasta with a thick sauce made from tomatoes, tomato paste, water, bell peppers, onions, garlic, mushrooms, Italian spices. In addition, have a green salad, wholegrain bread and fresh fruit for dessert.

Dr. Whitaker then suggests, "Other tasteful and healthy dishes are potatoes (mashed, baked or boiled) cooked with spices and sometimes fruit for a sweeter flavor, various bean casseroles and soups, sweet potatoes and, one of my favorites, succotash, which is a mixture of fresh kernel corn and lima beans.

"Be sure to diversify your foods, choosing some from each vegetable class. This insures an optimum balance of amino acids and adds zest to meals.

"Of all the tricks runners use to improve performance, the most important one is what they eat. Keep that in mind next time you're offered a Western omelet, and eat the bread instead."[78]

Eating Program (Meatless) for Fitness Buffs

If you consume animal products during a fitness program, you need not be concerned about a proper balance of amino acids. But if you prefer meatless fare, a little planning can do much to combine amino acids with fitness to create better health.

The Runner, in one of its Sportsmedicine columns, had this suggestion: "Runners who don't eat meat have to make sure they get enough calories, and learn to complement proteins. The protein in vegetables or grains is often deficient in one or more essential amino acids, but by combining different foods in one meal, you can easily obtain complete protein. Mixing grains with seeds or with legumes (such as beans, peas or lentils) provides complete protein. A little bit of complete protein, as found in milk or cheese, can help your body utilize the incomplete protein found in vegetable products as well."[79]

12 YOUR AMINO ACID COOKBOOK

TO SATISFY all tastes, this chapter offers a cross-section of different types of high amino acid recipes—meat, meatless and a combination of the two. They are designed to give you complete protein along with delicious taste, a combination that can't be beat!

DRY LEGUMES

These include beans, peas and lentils, to name a few. Combined with small amounts of meat, poultry, fish or cereal grains in stews, casseroles and soups, they make low-cost, amino acid-rich meals possible. Legumes are born mixers as well as meat extenders. They can be mixed with other vegetables, used to add dash to salads, or served as a side dish to complement meat at any meal. In short, they can be mixed with other foods to make a complete protein, or make a little complete protein go a long way.

Beans

Black beans have a mild, almost sweet taste. They are most often used in thick soups and in Oriental and Mediterranean dishes. You might try them with brown rice or cheese.

Blackeye peas (or cow peas) are small, oval-shaped and creamy-white with a black spot on one side. They are used primarily as a main dish vegetable and, of course, with a marrow bone, but try them with chicken or turkey.

Garbanzo beans (chickpeas) are nut-flavored. Can be used as a main dish vegetable. Combine with sesame seeds for a complete protein.

Kidney beans are large, white to red in color and firm-skinned, with a tender, sweet interior. Ever-popular in chili con carne, they add zest to salads and Mexican dishes.

Lima beans are an excellent main-dish vegetable. They are exceptionally good with chicken wings and curry and in casseroles. Try them with lean meats and cheeses, too.

Navy beans is a broad term for white beans and includes Great Northern and pea beans. Oval and white, these beans are a favorite for home-baked beans, soups, casseroles and salads. They hold their shape even when cooked tender.

Pinto beans are related to kidney beans. They are beige and have brown speckles, but cook to pink color. Most often found in chili and salads.

Peas

Whole or split, green or yellow, peas are a good source of amino acids along with iron, potassium and thiamine. They mix well with other foods. Serve them buttered to accompany a main dish; mashed and seasoned to add a flashy touch to any meal; and pureed in dips, patties, croquettes and even pudding.

Lentils

Low in fat, an excellent source of amino acids and of iron and potassium; they also contribute calcium and phosphorus. Try lentils with nuts, or as a replacement for noodles, potatoes, beans, brown rice and meat in many dishes. Be sure to try lentil soup.

Tofu

An amino acid food derived from soybeans. It is whitish and usually found in a plastic container in the produce section of your food market. Tofu absorbs the flavor of the foods with which it is prepared, so it mixes easily with any spice or dish.

Tofu, or soybean curd, is very versatile. It can be used in soups, casseroles, salads, stir-fry dishes, dips and spreads—the uses are almost endless.

Tofu is not a "complete" protein, as it has a limited amount of the essential sulfur-containing amino acid methionine; including brown rice with tofu will provide enough of this nutrient. Tofu is also low in saturated fat and calories and free of cholesterol.

Cheesy Eggplant Casserole

1 cup uncooked brown rice
1 medium eggplant, sliced in ½-inch thick rounds
2 tablespoons vegetable oil
1 small onion, chopped
1 clove garlic, minced
½ pound mushrooms, sliced
1 8-oz. can tomato sauce
1 teaspoon oregano
½ teaspoon basil
12 oz. lowfat (1%) cottage cheese
¼ cup skim milk
2 oz. part skim mozzarella cheese, shredded

Cook brown rice. Steam eggplant slices just until tender. Sauté onion, garlic and mushrooms in oil until brown. Add tomato sauce and herbs, scraping up drippings.

Simmer for a few minutes. Mix cottage cheese and milk. In an oiled baking dish, layer cooked brown rice, eggplant, cottage cheese and sauce. Repeat, ending with sauce. Top with mozzarella. Bake at 350°F for 30 minutes. Serves 6.

Macaroni and Cottage Cheese Casserole

- 2 cups wholegrain macaroni
- 1 medium onion, chopped
- ½ lb. mushrooms, sliced
- ⅔ cup sunflower seeds
- 2 tablespoons oil
- 1½ cups (1% fat) cottage cheese
- ¼ cup skim milk
- 1¼ cups yogurt
- Dash cayenne
- 2 tablespoons parsley flakes
- ½ cup bran or wheat germ (or combo)
- 2 tablespoons grated Parmesan

Sauté onion, mushrooms and sunflower seeds in oil. Cook macaroni and combine with mushroom mixture. Beat cottage cheese, yogurt, milk and seasonings together. Stir into macaroni mixture. Put in lightly greased casserole. Mix bran and Parmesan. Sprinkle on top. Bake at 350° F for 30 minutes. Serves 6.

Kasha and Bows

- 1 tablespoon polyunsaturated oil
- 1 large onion, chopped
- 1 cup of kasha (buckwheat groats)
- ½ cup of vegetable or chicken broth
- 1 cup of water
- ¼ teaspoon pepper
- 4 oz. flat or bow noodles

Sauté onion in oil until soft. Stir in kasha and cook a few minutes more. Add broth, water, pepper. Cover tightly. Reduce heat and simmer for 15 minutes or until all the water is absorbed. Meanwhile, cook the noodles or bows according to package directions. Drain noodles and stir into cooked kasha. Serve with a green salad. Serves 8.

Bean and Cheese Enchiladas

1 tablespoon polyunsaturated oil
⅓ cup chopped green pepper
½ cup chopped onion
⅛ teaspoon chili powder
¼ teaspoon rosemary
1 green chili pepper, seeded and finely chopped
1 lb. can tomatoes
6 oz. can tomato paste
15 oz. can kidney beans, drained and mashed
½ cup lowfat cottage cheese
¼ teaspoon pepper
2 oz. Monterey jack cheese, grated
8 frozen tortillas

In skillet, sauté green pepper and onion in oil; add chili powder, rosemary and chili pepper and continue to sauté for five minutes. Remove from heat and set aside. In another skillet, combine tomatoes, tomato paste, kidney beans and steam for seven to ten minutes. Do not boil! Combine sautéed foods from first skillet with the tomato-bean mixture from second skillet. Add cottage cheese, pepper, Monterey jack cheese. Insert in the frozen tortillas that have been partially thawed. Bake in 375° F oven for 15 minutes. Serves 8.

Cottage Cheese Lasagne

1 lb. can tomatoes
15 oz. can tomato sauce
1½ tablespoons oregano
1 bay leaf
¼ teaspoon fennel
1 teaspoon basil
8 oz. part skim mozzarella
1 large onion, chopped
2 cloves minced garlic
1 tablespoon polyunsaturated oil
1 lb. lean ground beef
½ lb. lasagne noodles
12 oz. carton 1% milkfat cottage cheese
1 oz. grated Parmesan

Simmer tomatoes, tomato sauce and seasonings while preparing meat. Sauté onions and garlic in oil until golden. Add meat and brown. Drain off fat. Add tomato sauce and simmer for an hour. Prepare lasagne noodles. In an 8 inch square dish start with a thin layer of sauce, then a layer of noodles, then cottage cheese. Cover with slices of mozzarella. Continue layering, ending with sauce. Sprinkle with parmesan. Bake at 350°F for 50 minutes or until lightly browned and bubbly. Serve with broccoli vinaigrette, Italian bread and fresh melon. Serves 8.

Golden Grains with Herbs

1½ cups uncooked oats
1 egg, beaten
3 tablespoons butter, melted
¾ cup chicken, beef or vegetable broth
2 tablespoons dried parsley flakes
½ teaspoon oregano leaves
½ teaspoon basil leaves

Combine oats and egg in medium-sized bowl; mix until oats are thoroughly coated. Add oat mixture to butter in 10 to 12-inch skillet. Cook over medium heat, stirring constantly, 3 to 5 minutes or until oats are dry, separated and lightly browned. Add remaining ingredients; continue cooking, stirring occasionally, 2 to 3 minutes or until liquid evaporates. Serve in place of rice or pasta. Serves 4.

Stuffed Pasta with Wheat Germ

8 manicotti tubes, wholegrain
½ cup finely chopped onion
2 tablespoons butter or margarine
2 cans (15 oz. each) tomato sauce
⅛ teaspoon garlic powder
⅔ cup wheat germ-bran combo
¼ cup grated Parmesan cheese
1 teaspoon oregano leaves, crushed
½ teaspoon basil leaves, crushed
¼ teaspoon pepper
6 oz. Monterey jack cheese, sliced
1 tablespoon minced parsley

Cook manicotti tubes according to package directions; drain and set aside. Sauté onion in butter until tender. Stir in tomato sauce and garlic powder. Set aside. Brown beef. Remove from heat. Add wheat germ-bran, Parmesan cheese, oregano, basil, pepper and ¾ cup sauce to beef. Mix well. Fill manicotti tubes with meat mixture. Add remaining meat mixture to sauce. Pour half of sauce into baking dish. Place stuffed manicotti on sauce in single layer. Top with Monterey jack cheese. Pour remaining sauce around outside edge. Cover. Bake at 425°F for 25 minutes until thoroughly heated. Sprinkle with parsley. Serves 4.

Kidney Bean and Corn Chili

1 small onion, chopped
1 clove garlic, minced
½ cup green pepper, chopped
2 tablespoons vegetable oil
2 cups frozen or fresh corn
3 cups cooked kidney beans
1 cup water
1½ tablespoons tomato paste
½ to 1 teaspoon chili powder
¼ teaspoon cumin
1 teaspoon oregano

Sauté onion and garlic in oil. Brown green pepper lightly. Add liquid, corn and tomato paste. Mash 1 cup beans and then add all the beans to the mixture. Add seasonings. Bring to boil. Simmer for ½ hour until thick, not watery. Serves 4.

Meatless Lentil Curry

4 onions, chopped
4 cloves garlic, minced
1 cup brown rice, uncooked
4 tablespoons oil
2 teaspoons curry powder
8 cups water
4 tablespoons lemon juice
2 cups dry lentils
2 10-oz. packages frozen spinach
1 cup plain yogurt

Sauté onions, garlic and brown rice in oil for 3 to 4 minutes. Add curry powder and stir through. Add water, lemon juice and lentils. Stir and simmer covered until tender (about 30 minutes). Add spinach. Cook another 15 minutes or until spinach is thawed and heated through. Stir to mix. Serve with about 2 tablespoons yogurt topping for each serving. Serves 6.

Cheesy-Potato Casserole

2 cups 1% fat cottage cheese
2 tablespoons oil
1 small onion, chopped
2 tablespoons wholegrain flour
1 teaspoon parsley flakes
1 teaspoon thyme
Pepper, as desired
6 medium potatoes, cooked and sliced
⅓ cup skim milk
½ cup bran or wheat germ or combo
2 tablespoons finely grated Parmesan cheese
2 tablespoons butter

Boil potatoes in scrubbed skins until tender. Preheat oven to 350°F. Grease a 1½ quart casserole. Beat cottage cheese and 2 tablespoons oil with a rotary beater until fluffy. Mix in chopped onion. Mix flour and seasonings. Place a layer of sliced potatoes in casserole. Cover with a layer of cottage cheese. Sprinkle with seasoned flour. Repeat layers ending with a layer of potatoes. Pour milk over potato-cheese layers. Mix grains with Parmesan cheese. Sprinkle over potatoes. Dot with butter. Bake 30 minutes. Serves 6.

Tabouli
(Middle-Eastern Bulgur Salad)

2 cups bulgur (or cracked wheat)
2 cups boiling water
1½ cups parsley, chopped
¾ cup mint leaves, chopped (or ½ cup parsley and 3 tablespoons dried mint)
1 bunch scallions, chopped—including green part
4 tomatoes, chopped
Romaine lettuce leaves
Dressing:
¼ cup polyunsaturated oil
¾ cup lemon juice
2 cloves garlic, minced
Dash hot pepper sauce
⅛ teaspoon freshly ground pepper

Pour boiling water over bulgur. Let stand while preparing vegetables. When wheat is soft and no longer hot, stir in vegetables (except lettuce). Mix dressing and stir into salad. Chill overnight. Surround it with lettuce leaves and serve. Serves 6 to 8.

Tuna Noodle Casserole

- 2 6½ oz. cans tuna
- 1 cup celery, chopped
- ½ cup onion, chopped
- 1 1-lb. can tomatoes
- 1 6-oz. can tomato paste
- ½ cup water
- ½ teaspoon garlic powder
- 1 teaspoon oregano
- ½ teaspoon basil
- ¼ teaspoon fennel seed (optional)
- 8 oz. wholegrain noodles
- 2 tablespoons grated Parmesan cheese

Drain tuna. Use 3 tablespoons of drained oil (if not soybean oil substitute a polyunsaturated oil) to cook celery and onion until soft. Add tomatoes, tomato paste, water, garlic powder, oregano, basil and fennel. Cover and simmer 20 minutes. Meanwhile cook noodles. Lightly grease 2 quart casserole. Layer in half of noodles. Top with half of tuna mixture and 1 tablespoon cheese. Repeat with remaining ingredients. Bake at 350°F for 30 minutes. Serves 6.

Strawberry-Pineapple Frozen Yogurt

- 1 envelope (1 tablespoon) unflavored gelatin
- ¼ cup cold water
- 2 cups plain low-fat yogurt
- 1 cup mashed or pureed fresh strawberries
- 1 cup crushed pineapple, juice packed, drained

Add cold water to small saucepan. Sprinkle gelatin over water and wait 5 minutes. Heat gelatin over low heat until dissolved. Stir well until mixed, then let cool. Stir in yogurt. Refrigerate in shallow dish until somewhat thickened (about 45 minutes). Add fruit and whip until light and airy (about 2 minutes, electric mixer). Gently pour into freezer tray (without dividers) or cupcake papers and freeze until firm, about 2 hours. Let stand at room temperature about 10 minutes before serving. You can use this recipe with other fresh fruits as well as fruits canned in fruit juices. Serves 8.

Light and Lemony Foam Pudding

1 tablespoon unflavored gelatin
1 tablespoon cold water
1 cup boiling water
½ cup cold water
¼ cup lemon juice
½ cup non-fat dry milk
½ cup ice water
¼ teaspoon lemon juice
1 teaspoon grated lemon rind

Soften gelatin in 1 tablespoon of cold water. Add boiling water to dissolve. Add ½ cup of cold water, ¼ cup lemon juice and rind. Then chill until very thick—about 90 minutes. Chill a deep mixing bowl and beaters. Add dry milk, ice water and ¼ teaspoon lemon juice to bowl. Beat until fluffy, then chill. Break up lemon mixture with fork and add to whipped milk mixture. Beat well with electric mixer until pudding is fluffy but not too soft. Chill until firm. Serves 6.

Fluffy Yogurt Gelatin

1 small package fruit-flavored gelatin
1 cup plain low-fat yogurt

Prepare fruit-flavored gelatin according to package directions. When gelatin has started to set, whip with beater until light and fluffy. Stir in yogurt until thoroughly combined with gelatin and place in refrigerator until set. Serves 6.

Wholewheat Rye Crackers

1½ cups wholewheat flour
¼ cup rye flour
¼ cup sesame seeds
1 cup polyunsaturated oil
½ cup water as needed

Mix flours and seeds together; add oil and blend well. Add enough water to make a soft dough. Form dough into a ball; then roll it to ⅛ inch thick on an unoiled baking sheet (or between two sheets of wax paper). Score with a knife to form rectangles or diamonds. Bake at 350°F about 20–30 minutes until the crackers are crisp and golden. Break apart along score lines. Makes 200 1-inch-square crackers.

Chewy Oatmeal Cookies

⅜ cup butter
⅛ cup honey
1 egg, slightly beaten
1½ teaspoons vanilla
½ cup wholewheat flour
¾ teaspoon baking powder
1 cup wheat germ
2 cups rolled oats
½ cup sun-dried raisins
¼ cup sunflower seeds, toasted
1 tablespoon water

Cream together butter and honey. Add egg and vanilla. Beat well. Stir together remaining dry ingredients and combine with creamed mixture and water. Stir until well blended. Drop by teaspoonfuls onto greased cookie sheet. Flatten slightly. Bake in preheated 375°F oven 9–12 minutes. Makes about 6 dozen 1-inch cookies.

Pecan Oatmeal Chewies

¼ cup butter
¼ cup honey
⅓ cup wheat germ
¼ teaspoon ground cinnamon
1½ cups old-fashioned oats, uncooked
1 cup chopped sun-dried apricots
⅓ cup coarsely chopped pecans
¼ cup dark seedless sun-dried raisins

Combine butter and honey in a saucepan. Bring to a boil over medium high heat, stirring constantly. Reduce heat and simmer 4 minutes, stirring constantly. Remove from heat. Mix in wheat germ and cinnamon. Stir in oats, apricots, pecans and raisins. Drop mixture by level tablespoonfuls onto wax paper, shaping into a mound. Chill until firm. Keep refrigerated in a tightly covered container. Makes 3 dozen.

Almond Nutballs

1 cup butter
¼ cup honey
1 cup almonds, ground
½ teaspoon vanilla extract
¼ teaspoon lemon extract
¼ teaspoon almond extract
2 cups unsifted wholegrain flour

Cream butter and honey until light and fluffy. Mix in almonds, vanilla, lemon and almond extracts until well blended. Stir in flour until well mixed. Shape mixture into 1-inch balls and place on ungreased baking sheets. Bake at 325°F until very lightly browned, about 18 minutes. Remove from sheets and cool on wire racks. Makes 4 dozen.

Muesli

2 cups uncooked oats
1¼ cups milk
¾ cup raisins, chopped prunes or chopped dried apricots
½ cup orange juice
⅓ cup wheat germ, bran or chopped nuts (or combination)
¼ cup honey

Combine all ingredients; mix well. Cover; refrigerate overnight or at least 8 hours. Serve with milk or cream as desired. Serves 6.

RECIPES FOR VEGETARIANS

Kidney Bean Tacos

(a) Sauce
6 medium tomatoes, chopped
1 cup onions, finely chopped
½ teaspoon garlic, minced
2 teaspoons dried oregano
1 teaspoon honey
1 teaspoon salt substitute
1 cup red wine vinegar

Combine ingredients in a bowl. Mix thoroughly and set aside.

(b) Beans

3 cups cooked kidney beans (directions on package)
½ teaspoon chili powder, or to taste
Pinch cayenne, or to taste
½ teaspoon salt subsitute, or to taste
1 teaspoon oil
½ cup chopped onion
2 garlic cloves, minced
2 medium tomatoes, chopped, or ⅔ cup canned

Combine the cooked kidney beans with the chili powder, cayenne, salt substitute. In a large frying pan, sauté the onions and garlic in the oil. Add tomatoes, cook for 3 minutes. Mash the beans and add them ¼ cup at a time to the onion and tomato mixture. Cook for 10 more minutes. Cover pan to keep the beans warm.

(c) Tortillas

2 cups cornmeal flour
1 cup water

Combine cornmeal and water; knead to blend well, adding a little more water, if necessary, to hold the dough together. Shape into 12 balls. Roll out or press each ball between 2 sheets of waxed paper or pat by hand to form a 6-inch circle. Bake on a hot lightly greased griddle until lightly browned (a minute or two on each side). Tortillas should be soft and pliable. Spoon the bean filling on to the tortillas. Roll. Top with the sauce. Yield: 12 tortillas.

Tofu, Rice and Leafy Greens

Oil as needed
1¼ cup (½ lb.) tofu cut into 1-inch cubes
½ lb. leafy green vegetables torn into bite-size pieces (e.g., Chinese cabbage, watercress, spinach)
Sesame seeds to taste
2 cups cooked brown rice
Soy sauce, as needed (preferably salt-free)

Oil a large frying pan and sauté tofu cubes about 5 minutes. Push cubes to center of pan and spread torn greens around them. Sprinkle with sesame seeds and soy sauce. Cover and steam until wilted (about 3 minutes); do not overcook. Remover from heat. Drain excess liquid. Serve with rice. Serves 2 to 3.

Sesame Seed Crackers

1½ cups wholewheat flour
¼ cup soy flour
¼ cup sesame seeds
¾ teaspoon salt substitute
⅓ cup oil
½ cup water (as needed)

Stir flours, seeds and salt substitute together. Add oil and blend well. Add enough water to knead dough into a soft ball and roll to a thickness of ⅛ of an inch. Cut it into cracker shapes and place on an ungreased cookie sheet. Bake at 350°F for about 15 to 20 minutes or until crisp and golden. Yield: 3–4 dozen crackers.

Garbanzo and Sesame Seed Spread

⅔ cup dry garbanzo beans (chickpeas)
1 large onion, minced and sautéed in 1 tablespoon sesame oil
1-2 garlic cloves, minced
Juice of 2 lemons
1 tablespoon soy sauce (salt-free)
½ teaspoon salt substitute
¼ cup sesame butter (tahini)
½ cup roasted sesame seeds, ground

Cook garbanzo beans according to package directions until very tender. Puree, or thoroughly mash them in a small amount of their cooking water, adding the sautéed onion and minced garlic. When thoroughly blended, add the remaining ingredients and mix well. Chill. Serve as a dip or filling. Serves 8.

Peanut Sunflower Tacos

¾ cup raw peanuts, cooked
⅔ cup sunflower seeds, cooked with the raw peanuts
1 6-oz. can tomato paste
½ teaspoon cumin seeds
½ teaspoon dried crushed chili peppers
Pinch cayenne
2-3 cloves garlic
⅓ cup roasted sunflower seeds
⅓ cup roasted sesame seeds
Tomato slices
Chopped green onions and parsley
Chopped lettuce tossed with wine vinegar
8 tortillas, 8-inch size (see Kidney Bean Tacos recipe)

Mix the cooked peanuts and cooked sunflower seeds, tomato paste, cumin seeds, chili peppers, cayenne and garlic in a blender until smooth. Turn the mixture into a small saucepan and cook over low heat until very thick. Stir in the roasted sunflower and sesame seeds. Place the folded tortillas, filling and remaining ingredients on separate platters. Each person can assemble individual tacos. Serves 8.

Chili Garbanzos and Mixed Nuts

⅓ cup non-dairy margarine
½ teaspoon chili powder
Dash of salt substitute
2 cups garbanzos, cooked but not mushy
2 cups mixed nuts

Melt margarine in a saucepan. Stir in chili powder, salt substitute, garbanzos and nuts. Turn into a 10 × 15 × 1 inch baking pan. Bake at 400°F for about 15 minutes. Serve warm or cold. Serves 6 to 8.

Peanut and Sunflower Butter

¾ cup peanuts
1 cup sunflower seeds
¼ to ½ cup peanut or sunflower oil
Salt substitute to taste

Combine the peanuts and sunflower seeds in a blender, adding the oil slowly. Add salt substitute to taste, if desired.

Mixed Bean Salad with Crackers

1 cup cooked garbanzos (½ cup dry)
1 cup cooked kidney beans (½ cup dry)
1 cup cooked black beans (½ cup dry)
1 cup cooked string beans
¼ cup diced pimiento
¼ cup diced onion
2 tablespoons oil
1 tablespoon lemon juice
¼ teaspoon salt substitute
¼ teaspoon dried basil
Dark leafy greens
Crackers

Cook beans separately until they are tender but still firm. Drain well. Combine all the beans, cooked string beans and the rest of the ingredients. Toss well and refrigerate. Serve on bed of dark leafy greens with crackers. Serves 4 to 6.

Pancake Delight

2 cups buckwheat pancake mix
¼ teaspoon each caraway seeds, turmeric, curry, powder, allspice
⅛ teaspoon celery seeds
¼ teaspoon onion powder
2 cups buttermilk

Place pancake mix in a large bowl. Add spices and buttermilk. Stir lightly until well blended. Pour onto a lightly greased griddle or frying pan. Turn once. Yield: 16 medium pancakes.

Soy-Brown Rice Loaf

2 cups cooked mashed soy beans
1 cup cooked brown rice
1 cup milk
½ cup wholegrain bread crumbs
1 tablespoon oil
1 tablespoon powdered vegetable broth
2 tablespoons minced onion
Salt substitute as desired

Mix all ingredients well. Place in oiled loaf pan. Bake in a moderate oven (350°F) for 45 minutes. If desired, moisten top with tomato sauce. Serves 4.

Nutty Biscuits

2 cups wholewheat flour
1 teaspoon baking powder
1 cup chopped nuts
Pinch of salt substitute
1 tablespoon butter
Skimmed milk, enough for mixture

Make a firm paste with blended dry ingredients, shredded butter and skimmed milk. Blend well together on a floured board and roll out to ½ inch thickness. Cut into circles. Prick well. Bake at 350°F on a greased tin, for about 20 minutes. Yield: 18 medium biscuits.

Potato Kugel

6 medium raw potatoes
2–3 raw carrots
1 large onion
1 clove garlic, minced
2 eggs, beaten
3 tablespoons oil
2 teaspoons salt substitute
¼ cup wholegrain bread crumbs
¾ cup dry skim milk powder
Topping, if desired: 1 cup grated cheese

Grate potatoes, carrots and onion into a large bowl. Drain off the accumulated liquid. Stir in the remaining ingredients, adding the milk powder slowly to avoid lumps. Spread mixture on an oiled 7 × 7 inch pan and bake at about 350°F for about 45 minutes to 1 hour. Kugel is done when edges are brown and an inserted knife will test dry. If desired, add the grated cheese topping; let remain in oven 5 more minutes until cheese melts. Serves 8.

Brown Rice and Raisin Custard

2 eggs
¼ cup honey
¼ teaspoon nutmeg
¼ teaspoon salt substitute
½ teaspoon vanilla
2 cups skim milk, scalded
1 cup cooked brown rice
½ cups raisins, washed and drained

Beat eggs slightly. Add honey, nutmeg, salt substitute, vanilla. Beat until blended. Add scalded milk gradually, stirring constantly. Stir in brown rice and raisins. Pour into an oiled 4 to 6 cup baking dish. Place baking dish in a pan of hot water. Bake at 350°F for about one hour or until set. Serves 4 to 6.

Cheese Muffins

2 cups sifted wholegrain flour
3 teaspoons baking powder
1 teaspoon salt substitute
4 tablespoons butter
¾ cup grated cheese
1 egg
1 cup skim milk

Combine sifted flour, baking powder and salt substitute. Cut in the butter with two knives. Add the grated cheese. Add egg to the milk and beat slightly. Pour the egg-milk mixture into the dry ingredients and stir quickly, just enough to moisten them. Fill oiled muffin tins about ⅔ full. Bake at 400°F for 20 to 25 minutes. Yield: 12 medium muffins.

With the preceding guidelines and recipes, you should be able to provide a balanced amino acid pattern in your food program while adhering to your dietary preferences.

TABLE 1. Amino Acid Content of Foods per 100 Grams—Edible Portion*

Food Item	Nitrogen Conversion Factor	Protein Content Percent	Phenylalanine Mg.	Isoleucine Mg.	Leucine Mg.	Valine Mg.	Sulfur Containing: Methionine Mg.	Sulfur Containing: Cystine Mg.	Sulfur Containing: Total Mg.	Tryptophan Mg.	Threonine Mg.	Lysine Mg.	Tyrosine Mg.	Arginine Mg.	Histidine Mg.
Milk, Milk Products															
Fluid, whole	6.38	3.5	170	223	344	240	86	31	117	49	161	272	178	128	192
Canned, evap.															
unsweetened	6.38	7.0	340	447	688	481	171	63	234	99	323	545	357	256	185
Dried, non-fat	6.38	35.6	1,724	2.271	3,493	2,444	870	318	1,188	502	1,641	2,768	1,814	1,300	937
Cheese, Cheddar,															
processed	6.38	23.2	1,244	1,563	2,262	1,665	604	131	735	316	862	1,702	1,109	847	756
Cottage	6.38	17.0	917	989	1,826	978	469	147	616	179	794	1,428	917	802	549
Eggs, whole															
fresh or stored	6.25	12.8	739	850	1,126	950	401	299	700	211	637	819	551	840	307
Meat, Poultry, Fish															
Beef, chuck, med. fat	6.25	18.6	765	973	1,524	1,033	461	235	696	217	821	1,625	631	1,199	646
Hamburg, reg	6.25	16.0	658	837	1,311	888	397	202	599	187	707	1,398	543	1,032	556
Rib roast	6.25	17.4	715	910	1,425	590	432	220	652	203	768	1,520	590	1,122	604
Round	6.25	19.5	802	1,020	1,597	1,083	484	246	730	228	861	1,704	661	1,257	677
Rump	6.25	16.2	666	848	1,327	899	402	205	607	189	715	1,415	550	1,045	562
Lamb, med. fat	6.25														
Leg	6.25	18.0	732	933	1,394	887	432	236	668	233	824	1,457	625	1,172	501
Rib	6.25	14.9	606	772	1,154	734	358	195	553	193	682	1,206	517	970	415

Food Item	Nitrogen Conversion Factor	Protein Content Percent	Phenylalanine Mg.	Isoleucine Mg.	Leucine Mg.	Valine Mg.	Sulfur Containing: Methionine Mg.	Sulfur Containing: Cystine Mg.	Sulfur Containing: Total Mg.	Tryptophan Mg.	Threonine Mg.	Lysine Mg.	Tyrosine Mg.	Arginine Mg.	Histidine Mg.
Pork, fresh, med. fat	6.25														
Ham	6.25	15.2	598	781	1,119	790	379	178	557	197	705	1,248	542	931	525
Loin	6.25	16.4	646	842	1,207	853	409	192	601	213	761	1,346	585	1,005	567
Pork, cured	6.25														
Bacon, med. fat	6.25	9.1	434	399	728	434	141	106	247	95	306	587	234	622	246
Ham	6.25	16.9	646	841	1,306	879	411	273	684	162	692	1,420	652	1,068	544
Luncheon meat, canned, spiced	6.25	14.9	570	741	1,151	775	362	241	603	143	610	1,252	879	942	479
Veal, med. fat	6.25	19.5	792	1,030	1,429	1,008	446	231	677	256	846	1,629	702	1,270	627
Round	6.25	19.5	792	1,030	1,429	1,008	446	231	677	256	846	1,629	702	1,270	627
Poultry, flesh only	6.25														
Chicken, fryer	6.25	20.6	811	1,088	1,490	1,012	537	277	814	250	877	1,810	725	1,302	593
Turkey	6.25	24.0	960	1,260	1,836	1,187	664	330	994	..	1,014	2,173	..	1,513	649
Fish	6.25														
Cod, fresh, raw	6.25	16.5	612	837	1,246	879	480	222	702	164	715	1,447	446	929	..
Haddock, raw	6.25	18.2	676	923	1,374	930	530	245	775	181	789	1,596	492	1,025	..
Halibut, raw	6.25	18.6	690	943	1,405	991	542	250	792	185	806	1,631	503	1,048	..
Salmon, Pacific, raw	6.25	17.4	646	883	1,314	927	507	234	741	173	754	1,526	470	980	..
Canned, sockeye or red	6.25	20.2	750	1,025	1,526	1,076	588	271	859	200	876	1,771	546	1,138	..
Meat Products	6.25														
Liver, calf	6.25	19.0	958	994	1,754	1,195	447	234	681	286	903	447	711	1,158	505
Bologna sausage	6.25	14.8	540	718	1,061	744	313	185	498	126	606	1,191	481	1,028	398
Frankfurters	6.25	14.2	518	688	1,018	713	300	177	477	120	582	1,143	461	986	382
Liverwurst	6.25	16.7	759	818	1,400	1,037	347	203	550	187	724	1,301	510	1,034	497

Legumes, dry and Nuts

Bean, red kidney, canned	6.25	5.7	315	324	490	346	57	57	114	53	247	423	220	343	162
Peanuts	5.46	26.9	1,557	1,266	1,872	1,532	271	463	734	340	828	1,099	1,104	3,296	749
Peanut Butter	5.46	26.1	1,510	1,228	1,816	1,487	263	449	712	330	803	1,066	1,071	3,198	727
Pecans	5.30	9.4	564	553	773	525	153	216	369	138	389	435	316	1,185	273
Walnuts	5.30	15.0	767	767	1,228	974	306	320	626	175	589	441	583	2,287	405

Grains and Their Products

Bread, white															
4% milk solids	5.70	8.5	465	429	668	435	142	200	342	91	282	225	243	340	192
Cereal combinations															
Infant food,															
precooked mixed															
cereal & dry milk	6.25	19.4	543	..	..	..	310	137	447	118	..	273	447	447	233
Oat-corn-rye, puffed	5.83	14.5	933	841	1,368	900	388	234	622	172	545	343	622	776	326
Corn Products	6.25														
Corn grits	6.25	8.7	395	402	1,128	444	161	113	274	53	347	251	532	306	180
Corn meal, degermed	6.25	7.9	359	365	1,024	403	147	102	249	48	315	228	483	278	163
Cornflakes	6.25	8.1	354	306	1,047	386	135	152	287	52	275	154	283	231	226
Hominy	6.25	8.7	333	349	810	398	99	..	..	84	316	358	331	444	203
Oatmeal, rolled oats	5.83	14.2	758	733	1,065	845	209	309	518	183	470	521	524	935	261
Rice, white or converted	5.95	7.6	382	356	655	531	137	103	240	82	298	300	347	438	128
Rice, products															
flakes or puffed	5.95	5.9	286	..	..	..	..	44	..	46	..	56	124	137	137
Wheat products	5.70														
Farina	5.70	10.9	579	..	..	..	143	184	327	124	..	199	447	424	268
Flakes	5.70	10.8	478	496	891	572	127	191	318	121	356	360	311	559	231
Macaroni or Spaghetti	5.70	12.8	669	642	849	728	193	243	436	150	499	413	422	582	303
Noodles, made with															
egg	5.70	12.6	610	621	834	745	212	245	457	133	533	411	312	621	301
Shredded wheat	5.83	12.8	755	..	..	..	..	246	..	136	..	466	481	742	371

Food Item	Nitrogen Conversion Factor	Protein Content Percent	Phenylalanine Mg.	Isoleucine Mg.	Leucine Mg.	Valine Mg.	Sulfur Containing			Tryptophan Mg.	Threonine Mg.	Lysine Mg.	Tyrosine Mg.	Arginine Mg.	Histidine Mg.
							Methionine Mg.	Cystine Mg.	Total Mg.						
Fruits															
Bananas, ripe	6.25	1.2	..	..	..	..	11	..	..	18	..	55	31	..	..
Grapefruit	6.25	0.5	..	..	..	..	10	..	..	1	..	30	..	..	..
Muskmelon	6.35	0.6	..	..	..	..	2	..	..	1	..	15	..	..	..
Oranges or orange juice	6.25	0.9	..	..	..	..	2	..	..	3	..	22	..	..	..
Pineapple	6.25	0.4	..	..	..	..	1	..	..	5	..	9	..	..	..
Vegetables															
Asparagus, canned	6.25	1.9	60	69	83	92	27	..	..	23	57	89	..	106	31
Beans, snap, canned	6.25	1.0	24	45	58	48	14	10	24	14	38	52	21	42	19
Lima, canned	6.25	3.8	197	233	306	246	41	42	83	49	171	240	131	230	125
Beets, canned	6.25	0.9	15	29	31	28	3	..	..	8	19	48	..	16	12
Beet greens	6.25	2.0	116	84	129	101	34	..	..	24	76	108	..	83	26
Broccoli	6.25	3.3	119	126	163	170	50	..	..	37	122	147	..	192	63
Cabbage	6.25	1.4	30	40	57	43	13	28	41	11	39	66	30	105	25
Carrots, raw	6.25	1.2	42	46	65	56	10	29	39	10	43	52	20	41	17
Cauliflower	6.25	2.4	75	104	162	144	47	..	..	33	102	134	34	110	48
Celery	6.25	1.3	..	..	..	..	15	6	21	12	..	..	..	..	..
Corn, sweet, white or yellow, canned	6.25	2.0	112	74	220	125	39	33	72	12	82	74	67	94	52
Cucumber	6.25	0.7	..	..	..	..	8	..	..	14	..	..	..	..	..
Eggplant	6.25	1.1	48	56	68	65	6	..	..	10	38	30	..	37	19
Lettuce	6.25	1.2	..	..	..	..	4	..	..	..	..	70	..	..	..
Onions, mature	6.25	1.4	39	21	37	31	13	..	..	21	22	64	46	180	14
Peas, canned	6.25	3.4	131	156	212	139	27	37	64	28	125	160	83	302	55

Potatoes															
cooked or canned	6.25	1.7	75	75	85	91	21	16	37	18	67	91	30	84	24
Pumpkin................	6.25	1.2	32	44	63	45	11	..	..	16	28	58	16	43	19
Radishes	6.25	1.2	..	..	..	30	2	..	..	5	59	34	..	..	..
Spinach	6.25	2.3	99	107	176	126	39	46	85	37	102	142	73	116	49
Squash, summer.........	6.25	0.6	16	19	27	22	8	..	..	5	14	23	..	27	9
Tomatoes, all types	6.25	1.0	28	29	41	28	7	..	..	9	33	42	14	29	15
Turnips................	6.25	1.1	20	20	..	..	12	..	..	..	..	57	29	..	..

* Figures for the amino acid content of foods are taken from Orr, M. L., and Watt, B. K.: Amino Acid Content of Foods. Home Economics Research Report No. 4. Washington, U.S.D.A., 1957. Amino acid content is given in milligrams, using whole numbers, rather than in grams, using decimals. The order of listing the amino acids has been arranged for the convenience of dietitians dealing with inborn errors and metabolism.

TABLE 2.
AMINO ACID BALANCE
Grams Per 100 Grams of Essential Amino Acids in Each Food

	THREO-NINE	*VALINE*	*LEUCINE*	*ISO-LEUCINE*	*LYSINE*	*METHIO-NINE*	*PHENYL-ALANINE*	*TRYPTO-PHAN*	*PROTEIN SCORE*
Ideal Protein	11.1	13.9	19.4	11.1	15.3	9.7	16.7	2.8	100
Whole Wheat	8.9	13.5	20.4	10.0	8.7	12.3	22.9	3.3	56.9
Soy Beans	9.8	12.2	19.8	11.6	16.2	6.6	20.6	3.3	68.0
Cow's Milk	9.4	12.3	20.2	10.0	16.5	7.0	21.5	3.0	72.2
Amaranth grain	11.4	10.6	14.8	10.2	16.6	11.2	23.1	2.1	75.0

From laboratory analyses by Indiginous Foods Consultants, Inc. Ann Arbor, Michigan (sponsored by Rodale R & D)

TABLE 3.
COMPOSITION OF RAW GREENS:
Selected Nutrients in 100 Grams

	MOISTURE (%)	*PROTEIN (Gr.)*	*CALCIUM (Mg.)*	*PHOSPHORUS (Mg.)*	*IRON (Mg.)*	*POTASSIUM (Mg.)*	*VIT. A (I.U.)*	*THIAMINE (Mg.)*	*RIBOFLAVIN (Mg.)*	*NIACIN (Mg.)*	*ASCORBIC ACID (Mg.)*
Amaranth	86.9	3.5	267	67	3.9	411	6,100	.08	.16	1.4	80
Beet Greens	90.9	2.2	119	40	3.3	570	6,100	.10	.22	.4	30
Chard	91.1	2.4	88	39	3.2	550	6,500	0.06	0.17	0.5	32
Collards	85.3	4.8	250	82	1.5	450	9,300	0.16	0.31	1.7	152
Kale	87.5	4.2	179	73	2.2	378	8,900	—	—	—	125
Spinach	90.7	3.2	93	51	3.1	470	8,100	.10	.20	.6	51

From Composition of Foods, Handbook No. 8 USDA

TABLE 4.
PROTEIN CONTENT OF FOODS— OFFICIAL U.S. GOVERNMENT LISTINGS

FOOD	*AMOUNT*	*GRAMS PROTEIN*
Milk and Milk Products		
Milk, Cow:		
Whole or non-fat fluid	1 cup	8.5
Non-fat dry, instant	1 tablespoon	1.8
Buttermilk, cultured	1 cup	8.7
Milk, goat	1 cup	8.1
Milk, human	1 cup	3.4
Cheese:		
Cheddar	2 tablespoons	7.1
Cheddar, processed	2 tablespoons	6.6
Cottage	2 tablespoons	4.8
Cream	2 tablespoons	2.6
Eggs		
Whole, large	1 egg	6.4
Meat, Poultry and Fish, Raw		
Beef, medium fat, without bone	½ cup	20.6
Chicken, fryer, flesh only	½ cup	23.4
Fish	½ cup	20.6
Heart, beef	½ cup	19.2
Kidney, beef	½ cup	17.0
Lamb, leg, without bone	½ cup	20.4
Liver, beef	½ cup	22.3
Sausage:		
Bologna	2 tablespoons	4.2
Frankfurter	1/10 lb.	6.4
Turkey, flesh only	½ cup	27.2
Veal, round, boneless	½ cup	22.1

Mature Legumes and Their Products

Beans, common	2 tablespoons	6.1
Chickpeas, garbanzo	2 tablespoons	5.9
Cowpeas	2 tablespoons	6.5
Lentils	2 tablespoons	7.1
Lima beans	2 tablespoons	5.9
Peas	2 tablespoons	6.7
Soybeans	2 tablespoons	9.9
Soybean flour, low fat	1 cup	45.1
Soybean milk	½ cup	3.9

Seeds, Nuts and Their Products

Brazil nuts	2 tablespoons	4.1
Coconut, fresh	2 tablespoons	1.0
Cottonseed flour	2 tablespoons	12.0
Filberts	2 tablespoons	3.6
Peanuts	2 tablespoons	7.6
Peanut butter	1 tablespoon	4.2
Pecans	2 tablespoons	2.7
Sesame meal	2 tablespoons	9.5
Sunflower meal	2 tablespoons	11.2

Grains and Their Products

Barley	2 tablespoons	3.6
Bread (4% non-fat dry milk, flour basis)	1/20 lb or 1 slice	1.9
Buckwheat flour, dark	1 cup	11.5
Corn and soy grits	1 cup	9.0
Corn Products:		
Flakes	1 cup	2.0
Grits	1 cup	13.9
Meal, whole	1 cup	10.9
Oatmeal	1 cup	11.4
Pearl millet	2 tablespoons	3.2
Rice, white	1 cup	14.5
Rye flour, light	1 cup	7.5
Sorghum, grain	2 tablespoons	3.1

FOOD	*AMOUNT*	*GRAMS PROTEIN*
Wheat Products:		
Flakes	1 cup	3.8
Flour, wholegrain	1 cup	16.0
Flour, white	1 cup	11.6
Germ	1 cup	17.1
Macaroni, elbow	1 cup	15.7
Noodles	1 cup	9.2
Shredded wheat	2 tablespoons	2.9
Vegetables, Raw		
Beans, lima	4 tablespoons	4.3
Beans, snap	4 tablespoons	1.4
Cabbage	4 tablespoons	.8
Carrots	4 tablespoons	.7
Collards	4 tablespoons	2.2
Corn, sweet	4 tablespoons	2.1
Cowpeas	4 tablespoons	5.3
Kale	4 tablespoons	2.2
Okra	4 tablespoons	1.0
Peas, green	4 tablespoons	3.8
Potatoes	4 tablespoons	1.1
Spinach	4 tablespoons	1.3
Sweet potatoes	4 tablespoons	1.0
Turnip greens	4 tablespoons	1.6
Miscellaneous		
Gelatin	1 tablespoon	8.6
Yeast:		
Compressed	2 tablespoons	3.0
Brewer's, dried	1 tablespoon	3.0

Source: Leverton, Ruth M. *Food, The Yearbook of Agriculture*. United States Department of Agriculture, Washington, D.C., 1959, pp. 71–73.

REFERENCES

1. Ruth M. Leverton, *Food: 1959 Yearbook of Agriculture*, Washington, D.C.: U.S. Department of Agriculture, 57–63.
2. Food and Nutrition Board, National Academy of Sciences-National Research Council, Washington, D.C., 1980.
3. Malden C. Nesheim, Ph.D., *This Medicine Called Nutrition*, Englewood Cliffs, N.J.: CPC North America, 1979, 6, 7.
4. Edmund Sigurd Nasset, Ph.D., *Food and You* (New York: Barnes and Noble, Inc., 1958), 4.
5. Carlson Wade, "Protein Builds More Than Muscles," *Muscle Training Illustrated* (Vol. 3, No. 2, March–April, 1967), 17.
6. *Nutritive Value of Meat*, National Livestock and Meat Board, Chicago, 1976, 5, 8, 26.
7. John D. Palombo, M.L.S., M.S., and George L. Blackburn, M.D., *Contemporary Nutrition* (Milwaukee, Wisconsin: General Mills, Jan. 1980, Vol. 5, No. 1), 1.
8. Donald L. Donohugh, M.D., *The Middle Years* (New York: Berkeley Books, 1983), 49–50.
9. Ruth M. Leverton, *Food: 1959 Yearbook*, 64–70.
10. Adelle Davis, *Let's Eat Right to Keep Fit* (New York: Signet Books, 1970), 37–38.
11. Yoon Sang Cho-Chung, M.D., Ph.D., quoted in "Amino Acids, Your Health Defenders," *Prevention* magazine, Aug. 1982, 54–55.
12. Yoon Sang Cho-Chung, M.D., Ph.D., quoted in *Science Magazine*, October, 1981.
13. *Arzneimittel-Forschung*, Vol. 26, No. 4, 1976.

14. *Nature Magazine*, March 22, 1979.
15. *American Journal of Psychiatry*, May, 1980.
16. *Lancet*, Aug. 16, 1980.
17. Ryan Huxtable, Ph.D., quoted in "Amino Acids, Your Health Defenders," *Prevention* magazine, Aug. 1982, 56.
18. *Dermatologica*, Vol. 156, 1978.
19. W. R. Beisel, M.D., quoted in *American Journal of Clinical Nutrition*, Vol. 30, 1236, 1977.
20. Dr. N. S. Scrimshaw, quoted in *American Journal of Clinical Nutrition*, Vol. 30, 1536, 1977.
21. Ronald J. Amen, Ph.D., quoted in "The Miracle Food That Pampers Your Heart and Peps Up Your Life," *Prevention* magazine, March, 1983, 79–80.
22. *American Journal of Clinical Nutrition*, Vol. 27, 175, 1974
23. *Medical World News*, Vol. 19, No. 19, 1978.
24. Ruth M. Leverton, *Food: 1959 Yearbook*, 65–70.
25. Roslyn Alfin-Slater and Lilla Aftergood, *Nutrition for Today*, (Dubuque, Iowa, William C. Brown Company, 1973), 11–14.
26. Edmund Sigurd Nasset, Ph.D., *Food and You* (New York: Barnes and Noble, Inc., 1958), 39.
27. Adelle Davis, *Let's Eat Right*, 39–41.
28. Ensminger, Ensminger, Konlande and Robson, *Foods and Nutrition Encyclopedia* (Clovis, Calif.: Pegus Press), 1983, 62.
29. Leo Szilard, "On The Nature of the Aging Process," *Proceedings of the National Academy of Science*, 45:30–45, 1959.
30. *Ibid.*
31. Dr. Leslie Orgel, "The Maintenance of the Accuracy of Protein Synthesis and Its Relevance to Aging," *Proceedings of the National Academy of Science*, 49:517–521, 1963.
32. Donald L. Donohugh, M.D., *The Middle Years* (New York: Berkeley Books, 1983), 17–18.
33. Hans J. Kugler, Ph.D., *Slowing Down the Aging Process* (New York: Pyramid Books, 1973), 44–46.
34. Adelle Davis, *Let's Eat Right*, 34–35.
35. Adelle Davis, *Let's Get Well* (New York: Signet Books, 1972), 310–311.
36. *ConsumerViews* (New York: Citibank), Vol. 3, No. 12, December, 1972, 3.
37. *Medical World News*, Oct. 7, 1978.

38. William Philpott, M.D., Data Sheet, Philpott Medical Center, Oklahoma City, Okla., 1980.
39. Hans J. Kugler, Ph.D., *Slowing Down the Aging Process* (New York: Pyramid Books, 1973), 28–29.
40. Dr. Howard J. Curtis, *Symposium on Topics in the Biology of Aging* (Salk Institute for Biological Studies, 1965), Peter L. Krohn, editor, New York: Interscience Publications, 1966.
41. *Health News & Review* July–August, 1984, p. 14.
42. Hans J. Kugler, Ph.D., *Slowing Down the Aging Process*, 28–42.
43. E. Consolazio, et al., *American Journal of Clinical Nutrition*, 28:29, 1975.
44. Denham Harman, M.D., "Prolongation of Life: Role of Free Radical Reactions in Aging," *Journal of the American Geriatrics Society*, 17:721–735, 1969.
45. Alice Kuhn Schwartz, Ph.D., quoted in "Meals to Match Your Mood," *Prevention* magazine, Jan. 1981, 74–79.
46. "Amino Acids," New York: *Health Foods Retailing Magazine*, Vol. 48, No. 6, June, 1984, 30–31, 62–66.
47. Emrika Padus, *The Woman's Encyclopedia of Health and Natural Healing* (Emmaus, Pa.: Rodale Press), 1981, 212.
48. Augusta Askari, Ph.D., quoted in "Zinc, the Mineral for All Reasons," *Prevention* magazine, June, 1982, 71–72.
49. George Schwartz, M.D., *Food Power* (New York: McGraw-Hill Book Company, 1979).
50. *Nutrition Reports International*, Jan. 1978.
51. Sheldon V. Pollack, M.D., quoted in "Nutrients That Help Your Body Heal Itself," *Prevention* magazine, Aug. 1982, 90.
52. *Journal of Nutrition*, Jan. 1979, 91–97.
53. Gerald Moss, M.D., quoted in *Medical Tribune*, Jan. 24, 1979, R26.
54. Jose A. Yaryura-Tobias, M.D., quoted in "Brain Food—It Really Works," *Prevention* magazine, Apr., 1981, 138–142.
55. *Eat to Live* (Rockville, Md.: Wheat Industry Council, 1976), 17–18.
56. Malden C. Nesheim, Ph.D., *This Medicine Called Nutrition*, 13.
57. "Vegetarian Follows Star Guide to Good Eating" (New York: City Health Department, Jan., 1977).

58. Roger W. Miller, "There's Something to Be Said for Never Saying, 'Please Pass the Meat,'" *FDA Consumer*, Feb., 1981.
59. Johanna Dwyer, D.Sc., "Vegetarianism," *Contemporary Nutrition* (Milwaukee, Wis.: General Mills), June, 1979.
60. Ruth Kunz-Bircher, *Eating Your Way to Health* (New York: Penguin Books, 1972), 16, 17, 286–287.
61. *Nutrition and Health*, (New York: Columbia University, 1980), Vol. 2, No. 1, 3–5.
62. Vic Sussman, *The Vegetarian Alternative*, (Emmaus, Pa.: Rodale Press, 1978), 85–86.
63. Henrietta Fleck, *Introduction to Nutrition*, 2nd edition, (New York: Macmillan Publishing Co., 1971), 63.
64. *Recommended Dietary Allowances*, National Academy of Sciences, 1974, 43.
65. N. S. Scrimshaw, *American Journal of Clinical Nutrition*, 1977, Vol. 30, 1536.
66. B. R. Bistrian, et al., *Journal of the American Medical Association*, 1976, Vol. 235, 1567.
67. George L. Blackburn, M.D., and B. R. Bistrian, M.D., *Surgical Clinics of North America*, 1976, Vol. 56, 1195.
68. Press Release, Medical & Pharmaceutical Information Bureau, Inc., Series 18, No. 12, August 17, 1982.
69. Niels H. Lauersen, M.D., and Eileen Stukane, *PMS: Premenstrual Syndrome and You* (New York: Fireside Books, Inc., 1983), 122, 123.
70. Patricia Allen, M.D., and Denise Fortino, *Cycles* (New York: Pinnacle Books, 1983), 148, 78–79.
71. Leo Wollman, M.D., *Eating Your Way to a Better Sex Life: The Complete Guide to Sexual Nutrition* (New York: Pinnacle Books, 1983).
72. C. W. Whitmoyer, Sr., D.Sc., *Your Health Is What You Make It* (Smithtown, N.Y.: Exposition Press, Inc., 1972), 66–67.
73. Vic Sussman, *The Vegetarian Alternative*, 38–39.
74. Morris Notelovitz, M.D., *Stand Tall!* (Gainesville, Fla.: Triad Publishing Co., Inc., 1982), 21.
75. Patricia Allen, M.D., and Denise Fortino, *Cycles*, 103.
76. Robert C. Atkins, M.D., *Dr. Atkins' Nutrition Breakthrough* (New York: Bantam Books, 1981), 123.

77. Exercise Leaflet, Metropolitan Life Insurance Company, Stay Well series, 1979.
78. Julian Whitaker, M.D., press release, "How Much Protein Do Runners Need?" Undated.
79. *The Runner* magazine, February, 1984, 18.